Defending the Land of Four Quarters

Globalization, Environment, and Sustainable Development in the Americas

By Lawrence A. Herzog, Ph. D.
San Diego State University

cognella™
San Diego, CA

Bassim Hamadeh, CEO and Publisher
Christopher Foster, General Vice President
Michael Simpson, Vice President of Acquisitions
Jessica Knott, Managing Editor
Kevin Fahey, Cognella Marketing Manager
Jess Busch, Senior Graphic Designer
Seidy Cruz, Acquisitions Editor
Sarah Wheeler, Senior Project Editor
Stephanie Sandler, Licensing Associate

First published in the United States of America in 2013 by Cognella, Inc.

ISBN: 978-1-5165-5168-2

cognella™

www.cognella.com 800.200.3908

Contents

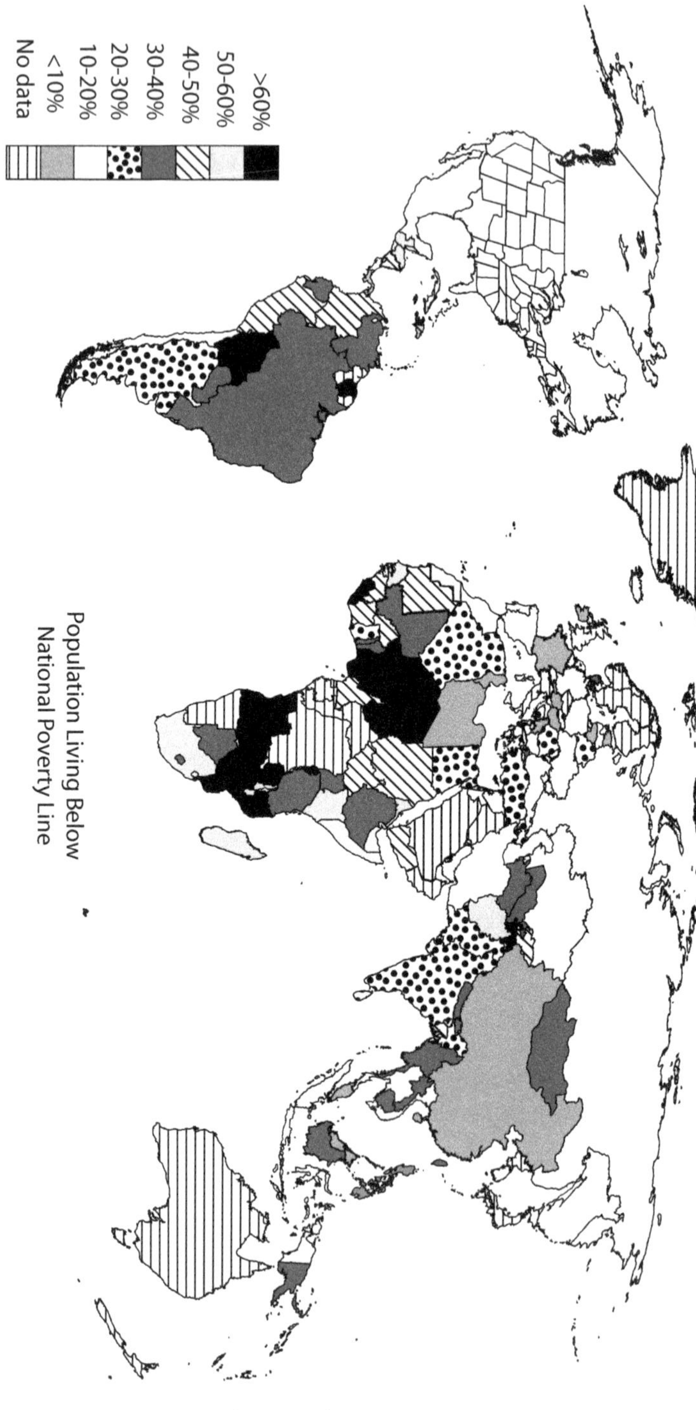

Fig. 1.1 Global measures of national poverty

Chapter 1

Introduction

During the last century, industrialization, urbanization, and, more recently, globalization, have all dramatically transformed the natural environment across the planet. Urban and industrial development are the cornerstones of what one might call a "western" model of national growth. I will argue in this book for a critical view of these western models of growth, noting their negative impacts on developing nations and entire regions, like Latin America. The critique will draw from examples across urban places, natural systems, and local cultures.

The ancient Inca civilization called its empire "Tahuantinsuyu" or "Land of the Four Quarters" because it stretched across the length and breadth of South America some 2,500 miles north and south from Colombia through the Andes mountains to Chile, and from the Atacama desert on the east coast to the Amazon rainforest in the west. At its peak, the Inca empire was the largest nation on earth, and it is considered the largest native American state that ever existed in the western hemisphere. Its sophisticated networks of highways, communication systems, vast Andean agriculture terraces, and finely chiseled stone cities all lie in ruins today, a metaphor for a world region that has been overrun by modern development, with devastating results.

Today, in what was once the "Land of the Four Quarters," we find a highly segmented region divided into dual societies of rich and poor. Inequality remains rampant. Economic systems south of the border have often not been favorable to local citizens; rather, they have

Fig. 1.2 Interior courtyard of colonial hacienda

favored export markets and international investors. Meanwhile, modern development has been destructive of local ecology and culture. This process, over time, has led to a geographic landscape of poverty and ecological decline—rural regions that can no longer sustain their peasant farmers, and mega-cities surrounded by rings of urban poor, or, as one author calls it, a "planet of slums."[1]

This book explores and reviews a set of themes associated with the transformation of social and physical environments in the Americas. It reviews some of the important streams of knowledge about sustainable development, with attention to both general and specific examples from Mexico to South America. The process of colonization and economic decline, referred to by scholars and international policy makers as "underdevelopment," remains a critical point of departure for thinking about the future economy and ecology of this region. It considers the degree to which these nations have become over-dependent on foreign investors and external economic forces, while sacrificing self-reliance and control over their own destinies.

These problems find their roots in history. Latin America was an indigenous society of farmers and crafts-people, sometimes living

1 See Mike Davis, *Planet of Slums* (New York: Verso, 2006).

within giant kingdoms like the Aztec, Inca, and Maya. Eventually, the indigenous cultures were conquered and colonized, with an Imperial system of economic life, city building, and social organization imposed by Spain and Portugal between the sixteenth and nineteenth centuries. The conquerors implemented the "western" or European model of economic development, which dramatically altered native culture, production, and the natural environment. Extractive economies—mining, agriculture, timber development, and fishing—were favored over sustainable agriculture and land utilization. Land was confiscated and placed into large-scale corporate estates or "haciendas" (*fazendas* in Portuguese). The colonists concentrated political and economic control over the land, leaving millions poor, landless, and outside the sphere of progress the minority enjoyed. The industrial revolution brought more foreign interests, railroads, factories, and other infrastructure in the nineteenth and twentieth centuries. Foreign investors came to earn huge profits from mining, agriculture, rubber production, and manufacturing. Cities grew; millions of rural peasants moved there to seek better lives, but many were underpaid or unemployed, and unable

Fig. 1.3 Favela on the edge of mega city, Rio de Janeiro, Brazil

to afford to live in standard apartments or houses. As a result, giant rings of slums grew around the edges of cities.

A Planet of Urban Slums and Degraded Nature

The growth of cities in developing nations during the twentieth century followed a similar pattern around the planet. It generally went through several stages: first, poverty created a push for rural peasants to move. Migrants then flowed toward small towns in neighboring regions to look for work; later, if local cities and towns did not have a large enough pool of jobs, migrants moved to larger cities, and ultimately, to the national capital, in search of industrial or other work that would allow them enough income to survive in a changing world. Those unable to find useful work often became international migrants or undocumented (illegal) migrants. This process of stepwise migration dominated the twentieth century.

The results of more than one hundred years of large-scale rural–urban migration are visible in the form of some of the largest, unplanned cities on the planet—giant "third world" cities where as much as half the population of the city lives in "spontaneous settlements" or squatter settlements. In addition, the cities suffer from problems ranging from lack of affordable housing, services, adequate diet, schools, and jobs to environmental degradation and crime. There are nearly 500 cities with more than a million inhabitants in Asia, Africa, and Latin America. By the early twenty-first century, less developed nations have even seen the evolution of "mega-cities," giant sprawling regions with more than ten million inhabitants—in Latin America, these include: Mexico City, Sao Paulo, Buenos Aires, and Rio de Janeiro.

While cities struggled with massive populations over the last five decades, both urban and rural environments suffered from the lack of ecological oversight. Factories were poorly regulated and generated air and water pollution. Land, water, and mineral resources were consumed without attention to their long-term preservation. Cattle ranches destroyed tropical rain forests. Croplands in floodplains were poorly managed. Insecticides, fertilizers, and other chemicals obliterated soils and fragile ecologies. Western or European strategies

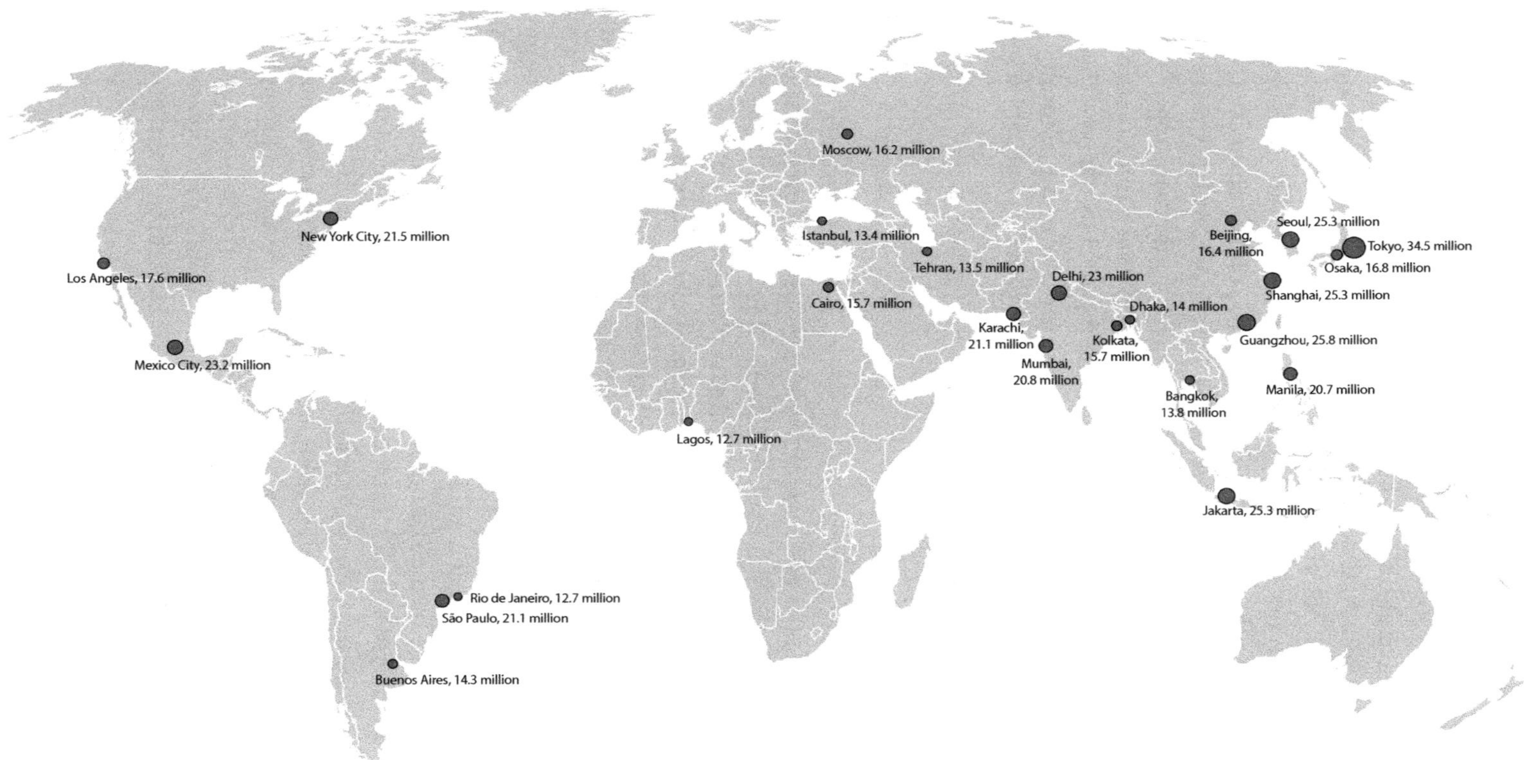

Fig. 1.4 Map of the thirty biggest mega cities in the World. The size of each circle indicates the number of inhabitants in millions.

emphasized "modernization" through increased technology, though this did not always yield positive outcomes. For example, agriculture expansion sometimes depleted soils, while insecticides killed fish stocks near agricultural land.

In Latin America, for example, timber production and ranching are wiping out the Amazon rainforest in Brazil; in Mexico, ranching, which once proliferated in the northern part of the nation, now dominates the tropical rainforest regions of the southeast. Over the last twenty-five years, more than one-fourth of Central America's rainforest has been turned into grass—to feed cattle that supply beef for US hamburger chains. Agro-industry throughout Latin America produces animal feed for export, while alienating the bio-regions where the agriculture is located. Waste and pesticides are dumped locally; workers are often underpaid or exploited and exposed to unhealthy, toxic work environments.

Inequality

Western styles of urbanization and industrialization have exacerbated inequality. Various explanations for this economic dualism have been offered by scholars. The simplest explanation is that when developing nations began to industrialize, some social groups were able to become part of the industrial, market economy—entrepreneurs, privileged families, factory managers, merchants, and so forth. Meanwhile, others—rural peasants, uneducated workers, and others were not able to advance upward and often remained underemployed or marginalized. This created, in effect, two sectors: a modern (formal) capitalist, industrial sector, and a traditional (informal) one based on subsistence/agriculture or menial urban jobs (street vending, dish washing, etc.). The informal sector is characterized by little or no job training, access to education, or marketing skills. Ultimately, in less developed nations, poor people who existed in traditional economies—agriculture, fishing, mining, etc.—were often further marginalized when either government or large corporations invested in those sectors of the economy, creating enterprises that would then outcompete rural peasants. No longer able to survive by selling their products, vast

Fig. 1.5 Street vendor, Bogotá, Colombia

rural–urban migration streams unfolded, generating giant informal economic sectors (low end, unskilled, unstable employment) in cities.

A second explanation, which follows logically from the first, is labeled "internal colonialism." Like economic dualism, it posits more deep-seated historical origins of inequality. Those origins include land concentrated in the hands of a small number of elite families and commerce and economic institutions also in the hands of privileged elites, largely excluding the poor from the political process. A third perspective, "dependency theory," has offered a structural explanation of underdevelopment which ascribes poverty and inequality in less developed nations to a "Core/periphery" metropolis/satellite view, in which it is argued that poverty is, in fact, systematically maintained by a process of social polarization that makes less advantaged nations dependent on the "core" countries.

This process plays out on both a global scale and a national or regional one. "Core" and "periphery" refer to where the power lies. The "core" is the colonizing power and its agents in the less developed country. The "periphery" refers to the people being colonized, and to those most marginalized by the dominance of the core. For example, in Latin America the "core" was represented by Spain and its institutions

Fig. 1.6 Squatter neighborhood near downtown Guayaquil, Ecuador

of domination: the King and royal family and the political agents who administered the colonial system—Viceroys, territorial governors, city officials. Within Latin American countries, the "core" was further represented by the large metropolis—usually also the national capital—where the political officials and economic leaders (plantation owners, investors, bankers, merchants, etc.) lived or had homes. The "periphery" was the rural areas where the poor lived, and later, the squatter settlements where poor migrants moved to try to improve their circumstances.

Globalization and Sustainable Development

The destruction of regional and national environments has been exacerbated by the process of globalization. Since around 1950, the planet has become increasingly "globalized," or interconnected, for a number of reasons: the globalization of finance within a hierarchy of global command centers or global cities, globalizing technologies (military, communication, transport), the increasing spread of cross-national markets, and international migration. Globalization has tended to exacerbate and intensify the dependency of less developed cultures on powerful and wealthy nations like the United States, Germany, Japan, and the United Kingdom. Further, by becoming enmeshed in the world trade/financial system, poorer nations are forced to focus on short-term economic "growth" strategies to pay off their international debt—rather than developing long-term sustainable economies that help prevent resource depletion and solidify regional and local productivity.

A set of "transnational practices" has been imposed on poorer nations—they include free trade agreements, international loans through organizations like the World Bank or the Inter-American bank, and consumer behaviors promoted through global media. The latter have been referred to by one scholar as a "culture-ideology of consumerism"—global advertising firms encouraging consumers everywhere to drink the same corporately produced soft drinks and coffee, consume fast food, or purchase generic products in malls or big box stores.[2] These practices cause further dependency on powerful corporate interests or nations and reduce local cultures' ability to produce and consume local products within sustainable bio-regions.

A central theme in this book is that the problems of poverty and underdevelopment in the Americas are partly the result of poor utilization of resources, combined with economic practices unfavorable to local cultures. The alternative model is "sustainable development"—economic and social growth that respects local and regional culture. How one creates this kind of economy is a question we will explore in this book.

Sustainable development is traditionally thought of as "meeting the needs of the present without compromising the ability of future generations to meet their own needs."[3] It brings to mind the oft quoted African proverb "We don't inherit the earth from our parents, we borrow it from our children."[4] Sustainable development thus raises the question of "carrying capacity"—how societies produce wealth and whether these behaviors represent the best use of regional resources.

2 Leslie Sklair, *The Sociology of the Global System* (Baltimore: Johns Hopkins University Press, 1991).

3 World Commission on Environment and Development, *Our Common Future* (New York: Norton, 1987).

4 Ibid.

Fig. 1.7 Global commerce: McDonald's in Buenos Aires, Argentina

Since the 1970s, this idea has been grounded in the theoretical idea that the environment is "socially constructed"; that is, humans are constantly imposing their will on the natural environment through their political and economic systems, settlement patterns, production of energy, transportation, and other lifestyle choices.[5]

In the past, it was assumed that all economic growth was good and that all forms of "modernization" (mining, mechanized agriculture, building factories, building cities, etc.) were something nations should aspire to. Sustainable development theory suggests that some forms of growth may be environmentally and culturally destructive. One example in Latin America is the colonial plantation or "hacienda" economy. While it successfully produced wealth for one segment of society, it was based on slave production and low wage labor. It destroyed

5 Michael Redclift, *Sustainable Development* (New York: Routledge, 1987).

planet green

indigenous land use patterns and social practices. It increased toxicity of watersheds due to pesticide and fertilizer uses and poor waste dumping practices. Developing nations, with the help of international organizations like the United Nations, or non-governmental agencies (NGO's), are now considering more "sustainable" development practices that better manage their ecosystems, conserve vital resources (forests, water systems, etc.), and allow these societies to become more self-reliant.

Fig. 2.1 Amazon rainforest near Madre de Dios river, southeast Peru

Chapter 2
Rethinking Development

The first task of adopting a sustainable approach to development in Latin America is to question what we mean by "development." During much of the twentieth century, it was assumed that development was a linear process—growth was projected to occur when labor and capital were applied to natural resources, leading to manufacturing and greater wealth. In traditional economic theory, growth was measured by Gross National Product (GNP) across various formal sectors: primary (agriculture, mining), secondary (manufacturing), and tertiary (services). Economists created models to measure goods, markets, price, and scarcity, determining optimal outputs with different combinations of these variables.

There are at least two major problems with this approach. First, it assumes that all resources are divisible and can be owned. But this covers only private goods. Much of a nation's resources are "public goods," like water, air, and forests, or, in short, the natural landscape. While air and water may seem obvious as public goods, a nation's flora and fauna (wildlife), especially on government-designated land (national, state, and local wilderness areas and open spaces), have too often been compromised in market-oriented economies. If nature is not protected—in the form of ecological planning and conservation—a nation's economy and social well-being may suffer in the long term. Also, GNP statistics and linear models of development don't take into account income distribution and inequality.

Fig. 2.2 The Atacama Desert bio region, Chile

A sustainable approach, therefore, revisits the question of "development." It argues that development must be understood as more than "economic growth" (the production of wealth), or "modernization" (buildings, roads, bridges, factories, office buildings, computer systems, etc.). Development must be viewed as a combination of productivity, social equity, and ecological preservation. Nature must become more central to our understanding of capitalism, production, exchange, and the meaning of value in a region's economy.[1]

"Eco-Development" and "Bio-Regions"

We can call this alternative approach "eco-development"—economic development tied specifically to cultural/ecological regions and local contexts, thus consistent with the potential of specific regions and their unique people, geography, and resources. This approach assures a more rational use of local resources, technologies that are appropriate to those regions, and the promotion of self-reliance. Eco-development should take place within ecologically defined units of territory that are often referred to as "bio-regions." A bio-region is an area created by

1 See Michael Redclift, *Sustainable Development* (New York: Routledge, 1995).

Fig. 2.3 Example of agro-industry: soybean plantation, state of Rio Grande do Sul, Brazil

natural systems and culture—for example, a valley, the watershed of a river, a desert basin, or a coastal port zone.

The concept of "carrying capacity" is also essential to an "eco-development" approach. "Carrying capacity" refers to the natural resource impacts of different human behaviors (economic practices) within a bio-region. We might ask the question: can a bio-region with a certain population profile absorb a given economic activity? For example, some high technology agriculture in sub-regions of Latin America, while it may be profitable, introduces damaging chemical (herbicides, fertilizers, etc.), which pollute the air and water and deplete soils. While short-term profits occur, there is destruction of the ecosystem needed to sustain growth over the long term.

In the past, eco-development approaches were overshadowed by government focus on international trade. Too often economic "growth" took precedence over resource conservation, renewable energy, and sustainability. This trend was exacerbated by the role of "foreign aid"—loans, grants, technical assistance, food aid, and development projects—which further forced nations to favor the creation of capital intensive projects that generated national profit to pay off foreign loans, or to buy goods needed for export-oriented growth. Meanwhile, resource sustainability and social needs were neglected.

At the micro or regional level, sustainable development is compromised by a phenomenon termed the "monetization" of local economies. "Middle men" and elite entrepreneurs often control key national economic sectors (large-scale agriculture, commerce, mining, etc.), while the poor and working classes tend to remain trapped in low-paying jobs. Meanwhile, national governments try to avoid a debt crisis by using public monies to invest in profitable industries tied to low wages. Not enough public funding is allocated to building sustainable and equitable economies that employ the masses at decent wage levels.

Agriculture is a good example of a local economic sector that has been globalized to the detriment of poor farmers all over Latin America. "Agro-industry" involves the production, processing, marketing, and distribution of food and fiber-products. These products acquire value depending on the global market. Because Latin America and other less developed regions of the world have become dependent on wealthy nations' consumers, they shift their agricultural policies to

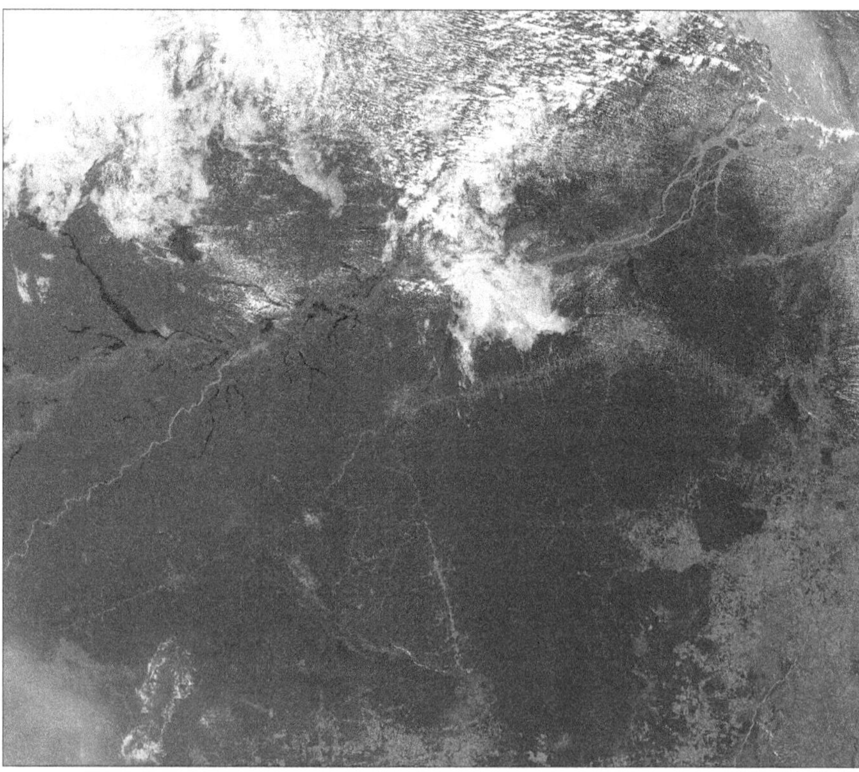

Fig. 2.4 Satellite image of the Amazon rainforest.

Fig. 2.5 Trans-Amazonian Highway at Pomba, Brazil

remain plugged in to the buyers' markets in those nations (in Europe, the US, Japan, etc.). In altering their farming practices to suit global markets, Latin American nations have ended up compromising the sustainability of their agricultural growing zones.

Fig. 2.6 Logging in the Amazon Rainforest

A Case Study of a Threatened Bio-Region: The Amazon Rainforest, Brazil

The Amazon Rain forest in Brazil constitutes about one-third of all rainforests of the world, some 2.3 million square miles. It generates 20 percent of the world's oxygen. Its main waterway, the Amazon River, contains 20 percent of the volume of all rivers on the planet. It may be the most ecologically diverse bio-region on Earth, with more than 56,000 species of plants and thousands of species of birds, amphibians, mammals, and reptiles. Its plant life contains natural remedies for some of the worst diseases plaguing humankind.

This bio-region holds value for Brazil and for the larger planet. Yet, in the last thirty years, corporate interests and the Brazilian government have exploited the region for short-term profit (mainly the sale of beef from cattle ranching, the sale of lumber from logging, and more recently the sale of soybeans), while threatening its long-term survival. Some 232,000 square miles of the Amazon have been deforested since 1978. It is an excellent case study to illustrate the problems and politics of unsustainable development.

There are a number of forces responsible for Amazon deforestation. They include:

1. Land clearing for cattle ranching. Cattle ranching is the leading cause of deforestation in the Brazilian Amazon. Cattle ranchers clear land to plant grasses that provide feed for herds of beef cattle. The world demand for Brazilian beef has made it very profitable as a short-term use of tropical lands in the Amazon. World demand grew during the 1980s and 1990s when Brazil devalued its currency, thus making its beef even more competitive in the world market. Road construction, including the Trans-Amazonian highway built several decades ago, opened forest lands and made shipping and packing of beef cheaper.
2. Colonization and subsistence agriculture. The Brazilian government has created programs for poor farmers to colonize jungle areas. Unfortunately, the government does not assist farmers in establishing ecologically sustainable best practices for farming. Peasant farmers use fires to clear land, then plant bananas, rice, maize, manioc, and palms, which grow well in the short term but

ultimately deplete soils of their nutrients, thus causing farmers to move on to other lands and repeat the same land clearing and inefficient agricultural practices.

3. Infrastructure. In the 1970s, to develop the rainforest, the federal government created the Trans-Amazonian Highway, a planned 2,000-mile highway system designed to open land for farmers, timber, and mining. The government spurred interest by offering subsidized land, a half-year's salary, and other perks. The highway was plagued with problems due to heavy rains, unstable soils, erosion, and poor farming yields for settlers. The highway has allowed more access and thus more deforestation as land is cleared for truck farms, cattle, and other production.
4. Commercial agriculture. Large scale agri-business arrived in the Amazon in the form of soybean cultivation. Brazilian scientists developed a new variety of soybean that grows well in the rainforest ecosystem. High demand for soybeans accelerated soybean production in the jungle.

Fig. 2.7 Tapping rubber

Fig. 2.8 Portrait of Chico Mendes

5. Logging. The biggest problem with logging is that, although there are laws in the Amazon controlling lumber production, the laws are not sufficiently enforced. There is a great deal of illegal logging in the Brazilian Amazon. Logging is closely tied with road

> building, so the more the government creates road access into the interior of the Amazon, the more loggers will get in to cut out trees. Those same roads also then encourage poor settlers to colonize the interior for temporary agriculture, access to building materials, fuel wood, and hunting game for food.

Taken together, these economic practices are wreaking havoc on one of the world's richest bio-regions, severely threatening its long-term survival and compromising what could be Brazil's richest source of long-term revenues.

Environmental Management and Social Movements: The Chico Mendes Story

Unfortunately, in developing nations, governments are not always the best environmental managers. Elected officials are strongly influenced by powerful corporate interests, as well as the need to pay off international debt. Their decisions regarding natural resources like the Amazon basin are strongly influenced by these two forces. International agencies can play an important role—the United Nations has recognized environmental issues but is sometimes sidetracked by other agendas. The United Nations Environmental Program is poorly funded and has no real jurisdiction over sovereign nations. Non-governmental organizations (NGOs) have stepped up to address environmental issues—for example, the World Wildlife Federation, Green Peace, and others. The World Conservation Union (IUCN, International Union for Conservation of Nature) has stated three ecological strategies that must be followed to make agriculture more sustainable: a) careful measurement of tradeoffs between crop management and cattle ranching; b) better crop management; c) protection of watershed forests. Mechanization and the globalization of agriculture are huge threats to preserving local farming.[2]

One hopeful trend is the recognition of "biosphere reserves." A biosphere reserve is an international conservation unit created by the

2 See IUCN, World Conservation Strategy, http://data.iucn.org/dbtw-wpd/edocs/WCS-004.pdf.

Fig. 2.9 Cloud Forest near Santa Elena, Costa Rica

United Nations Educational, Scientific, and Cultural Organization (UNESCO) under its Program on Man and the Biosphere (MAB). According to "The Statutory Framework of the World Network of Biosphere Reserves," biosphere reserves are created "to promote and demonstrate a balanced relationship between humans and the biosphere." Biosphere reserves serve in some ways as "living laboratories" for testing out and demonstrating integrated management of land, water, and biodiversity.[3] The Amazon Rainforest is one of several biosphere reserves in Brazil, though its designation is mainly for scientific purposes and does not imply political control by UNESCO or any other non-Brazil government entity.

While biosphere designations offer hope for regions like the Amazon Rainforest in the future, in the short term, local communities have often sought to take more direct and immediate actions. One such

3 UNESCO Man and Biosphere website http://www.unesco.org/new/en/natural-sciences/environment/ecological-sciences/biosphere-reserves/.

example is the case of a rubber tapper who became an environmental activist in the Amazon. His name was Chico Mendes. Mendes observed the decline in rubber prices in the 1960s, with landowners selling their properties to the highest bidder—typically cattle ranchers. Rubber tappers were being driven from their lands. In the 1970s, he joined the Amazon rubber tappers. They would march down logging trails and overrun forest clearance parties, disarming guards and attempting to convince the ranchers' employees not to continue. In many cases, they were successful at doing so, despite resistance from the ranchers.

Chico Mendes ultimately organized a social movement that advocated creating forest reserves that would be managed by traditional communities, which could cultivate locally sustainable products like rubber and Brazil nuts. His goal was to round up the rubber workers and form a union, and then challenge the forces that were causing deforestation—road pavers, cattle ranchers, and outside investors. Mendes became a hero to international environmentalists around the world, and, by 1987, he was in touch with the Environmental Defense Fund and the National Wildlife Federation. Mendes sought to convince the Inter-American Development Bank that its road project in his area would end in disaster, unless it took into consideration the preservation of the forest and the livelihoods of its inhabitants. He was successful, with the project first being postponed, and then, with his participation, renegotiated. He won two international environmental awards for this.

In 1988, Mendes launched a campaign to stop an Amazon rancher from logging an area that was a planned reserve. Mendes not only managed to stop the planned deforestation and create the reserve, but also gained a warrant for the rancher's arrest for a murder committed in another state. He delivered the warrant to the federal police, but it was never acted upon. One week after his 44th birthday, Chico Mendes was assassinated. His murder made international headlines, including the front page of the *New York Times*. Thanks in part to the international media attention surrounding the murder, the Chico Mendes Extractive Reserve was created in the area where he lived.[4] There are

---

4 For a full account of the Chico Mendes legacy, see Andrew Revkin, *The Burning Season: The Murder of Chico Mendes and the Fight for the Amazon Rain Forest* (New York: Collins, 1990)

more than twenty such reserves now, along the same lines as Mendes had proposed, covering more than eight million acres.

Ecological Management in Costa Rica

"Costa Rica touches the heart and mind, not through elegant boulevards, towering cathedrals, or an imposing place in history, but through its natural beauty..."[5]

Costa Rica offers a critical example of a natural resource-rich nation in Latin America that will benefit from a sustainable/bio-region-based approach to the protection of its valuable ecological resources. This 19,650 square mile nation has one of the more diverse ecological settings of any country in the western hemisphere—a mix of terrain, biota, and weather across Atlantic and Pacific coastal zones, tropical mountains, high altitude basins, rain forests, cloud forests, and volcanic peaks. It lies on an ecological land bridge between two continents—North and South America. It sits on the intersection of two tectonic plates beneath the earth's surface—the Cocos Plate (Pacific Ocean) and the Caribbean Plate (Atlantic side). The constant shifting of these plates has created both the mountainous topography in the nation and its sometimes active volcanoes and earthquake faults.

Biological diversity in Costa Rica is impressive: an estimated half million species of plants and animals, including 300,000 insects.[6] This has led the national government to create huge protected ecological zones, including twenty-seven national parks and fifty-eight wildlife refuges, as well as mangrove/wetland reserves, forest reserves, conservation regions, and protected zones covering nearly 25 percent of the national territory. These ecological reserves have been the catalyst for a booming eco-tourism industry. Visitors are drawn to the tropical highlands with their exotic wildlife, as well as to lowland rainforests. One of the more unusual eco-zones is the so called "cloud forest," which is a highland mist created by high altitude combined with wind and humidity, which provides an unusual constant source of moisture for

---

5 Ree S. Sheck, *Costa Rica: A Natural Destination* (Santa Fe: John Muir Publications, 1996)

6 See Costa Rica's InBio, Instituto Nacional de Biodiversidad at http://www.inbio.ac.cr/en/biod/bio_biodiver.htm

giant trees, mosses, and bromeliads, as well as exotic birds and insects. Costa Rica also has a valuable coral reef ecological reserve.

All of these ecological resources are fragile. As the nation's tourism industry has boomed, the very commodity that fuels that industry—the environment—is threatened by too many visitors, and by export markets that seek lumber, beef, or coffee. The biggest threat to wildlife survival in Costa Rica is deforestation. In 1950, 72 percent of the nation was covered by forest; by 1985, that number had dwindled to between 35 and 50 percent, according to various studies.[7] Much of the deforestation is the result of logging and clearing of forests for coffee growing or cattle grazing land. Many wildlife refuges are actually on privately owned land or lightly managed public lands that are difficult to manage.

One hopeful sign is the attention to Costa Rica's environment both from international NGOs and the national government. The government, for example, has bought up some vulnerable lands and put them in protected status. It is trying to plant trees in areas that were illegally deforested. Costa Rica has signed "carbon bond" agreements with nations like Norway, which has purchased 200,000 tones of "carbon bonds," an investment in protecting Costa Rica's forests by planting new trees. Eco-lodges and eco-tourism will serve as an incentive for the government to continue its conservation efforts, since ecological tourism has become such a huge part of the nation's economy.

7 See Monga Bay website http://rainforests.mongabay.com/20costarica.htm

Chapter 3

Sustainability, Nature, and Local Culture

Sustainability involves "paradigm shifts," or new ways of thinking about space, place, and nature. These shifts in how we think about the world can be thought of through a number of categories of ideas. This chapter reviews several of these.[1]

Time

In western culture, we tend to emphasize immediacy and instantaneity. Our technologies provide rapid access to information—on computers, cell phones, iPhones, or Blackberries, where we Google or surf the Internet, or on our TVs, where we channel surf. We have developed a sense of entitlement to information. These cyber-electronic machines feed our thirst for speed—they make us expect information to be accessed quickly; we are, therefore, becoming impatient with slowness and waiting.

This culture of immediacy spills over into the way we govern and manage space and our perception of the politics of the environment. In formal governance circles, there is a tendency to think and operate in short-term frameworks. For example, we make our urban and

1 Some of these points are raised in an excellent book—see Stephen Wheeler, *Planning for Sustainability* (New York: Routledge, 2004).

environmental plans fit into relatively brief time spans—five, ten, or twenty-year units. Meanwhile, our elected officials hold offices for even shorter time frames, from 1–4 years, with re-election for perhaps a similar amount of time.

Sustainability requires a different mode of thinking. It requires us to understand that the earth changes in long-term cycles—geological shifts, rivers, watersheds, and climate change all occur over massive periods of time. One of the problems of the debate on global climate change, or on the use of sustainable practices, is that politicians and voters are not accustomed to thinking in the long term. One reason for this is also that it requires us to put aside our selfish needs in the short term and consider the future of the planet, and of the next generations. In short, we need an environmental-historic approach. This will allow citizens to understand that, as an example, cleaning up the environment can takes decades. But not cleaning the environment, or not

Fig. 3.1 Earth evolves over long periods of time: uplifted rock formations, Big Bend River, Texas

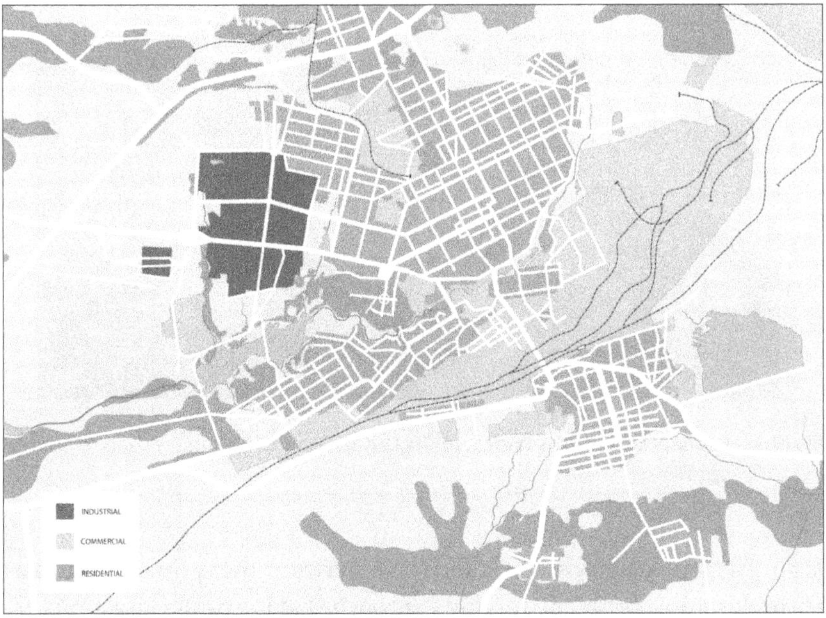

Fig. 3.2 An example of a typical city land use plan.

understanding the long-term dangers of pollution, could permanently scar the planet in ways that we might never recover from.

Holistic vs. Fragmented Knowledge

Another important distinction must be made between what might be termed "holistic" approaches to environmental planning, as opposed to "fragmented" approaches. In our western culture, we have a tendency to compartmentalize knowledge; we have become an overspecialized society. We train ourselves in very narrow disciplines—law, biology, marketing, etc., but often fail to understand the larger connections between things. Conceptually, this problem is identified as one of "fragmented knowledge." It is sometimes also referred to as "reductionism." We have become a culture of experts who know a lot about one thing, but not enough about larger interconnections. This tendency manifests itself in the environmental and urban planning fields—where we train specialists in oceanography, housing, or land use law, yet not enough

Fig. 3.3 Portland, Oregon is known for its Urban Growth Boundary used to control sprawl

scientists who can cross back and forth between disciplinary boundaries to develop a problem-solving approach that keeps the larger picture in mind. This larger picture is often referred to as "holistic knowledge."

An example of reductionism can be seen in Comprehensive Urban or Environmental Plans, or what are sometimes called Master Plans. These plans tend to be sub-divided into separate categories, or "General Plan Elements"—land use, open space, housing, urban design, transportation, etc. Policy decisions are then made within each of these General Plan Elements, often without recognizing the interconnections between them. Yet transportation planning must take into account land use, since transit systems are inherently tied to land use densities. Equally, an urban design element of any city or town is intimately connected to the location of open space, housing, and transportation infrastructure. In a sense, the idea of subdividing a General Plan or Master Plan into "elements" may no longer be the most useful way to prepare such a plan. Perhaps future plans will be divided into sections that address sustainable development strategies: equity, carrying capacity, ecological footprints, and so forth.

Limits

Another important dimension of sustainability is the recognition of limits. In western culture, we tend to make a linear connection between growth and progress. A 1972 book, *The Limits to Growth,*[2] tried to model the consequences of a rapidly expanding world population and finite resource supplies. It argued that if we continued to grow at current rates on this planet, there might not be enough resources to house and feed projected future populations. The book echoed concerns from Thomas Robert Malthus's 1798 book, *An Essay on the Principle of Population* (1798).

Western culture is deeply embedded in a period of heightened consumerism, in which citizens are constantly bombarded with subtle and direct messages that consumption is good, the more the better. Indeed, some have argued that western culture socializes citizens to believe that their self-worth is directly related to material acquisition.[3] Our theories of neo-classical economics are based on "growth is good" principles, reduced to concepts such as production, consumption, supply, and demand. The fields of economic planning and urban planning are all based on growth. Urban planning, for example, speaks to regulating growth, not how to limit growth or curb its impact on the environment.

Thus, to speak of limits is path breaking, even in the twenty-first century. Yet to ignore the concept of limits is to endanger the planet. Today, in some parts of the world, we are seeing the introduction of "urban growth boundaries" or lines that limit where growth takes place, and government sanctioned limits on water use (for example in desert regions) or on resource use (in national forests, in heavily fished seas, etc.).

---

2 See Donella H. Meadows, Dennis L. Meadows, Jørgen Randers, and William W. Behrens III), The Limits to Growth (New York: Universe Books, 1972).

3 This argument has been suggested in Leslie Sklair, *The Sociology of Global System* (Baltimore: Johns Hopkins University Press, 1990).

Place and Sustainability

The emphasis on specialized science and rationality also spills into how we study nature and place. We tend to reduce the natural environment to scientific categories (air, water, land, etc.), and then study it with generic land use models. But places are not merely neutral spaces to be broken down into scientific or quantifiable categories—they are unique geographic zones, with their distinct natural landscapes, histories, people, and local characters. World cultural regions must protect their uniqueness, in part, because it is how they can develop competitive economies—for example in creating tourism activities; in attracting investors, workers, and residents; and in building specialized economies around the mix of site attributes unique to each place.

One of the problems of protecting unique places is that global corporate interests often find it profitable to exploit either their natural resources (as in the Amazon rainforest) or their cultural resources for tourism development. In both cases , too much tourism development can destroy the very thing that tourists come to see. For example, too many cruise ships visiting Antarctica may compromise the very arctic environment visitors come to see. Overdevelopment along pristine coastlines may destroy the coral reefs and other ocean ecology that make these places attractive to tourists in the first place.

In many western societies, the thing that connects people to place is their community._And yet, there has been a well-documented decline in community life after 1950. The shift toward automobiles and sprawling suburbs has eroded pedestrian-scale neighborhoods, thus diminishing face-to-face contact between local residents. At the same time, technological "advances"—computers, cell phones, smart phones, and the Internet—further isolate individuals by allowing what were once social experiences (shopping, gathering information, meeting) to be replicated with machines. This is part of a larger phenomenon of privatization—as people moved to suburban homes in auto-oriented subdivisions, travelled by car to work and for daily errands, and as technologies enhanced doing those daily errands electronically, the need for personal interaction further diminished. Indeed, the fear of crime has even driven many citizens to live inside "gated" communities, thus further isolating people from authentic community experiences.

Fig. 3.4 This clothesline represents energy efficiency

Fig. 3.5 Electric clothes tumble dryers are not necessary in warm places on the planet

Meanwhile, it has also been documented that developers, politicians, and investors are creating more artificial spaces to encourage consumerism—shopping malls, festival marketplaces, and "themed" residential areas. The growth of simulated places in cities means a further loss of citizens feeling connected to place.

Place Destruction and the "Clothesline Dilemma"

One way to envision the problem we have with respecting the sacred nature of place and the environment can be expressed by what scholars term the "clothesline dilemma."[4] Imagine that you have just washed some clothes. You have two choices in drying them. We can call these choices Scenario 1 and Scenario 2.

Scenario 1. In scenario one, you want to dry clothes in a standard clothes-drying machine. There are at least nine steps taken when you choose this scenario. These steps would include:

- drill for oil in an oil-rich geographic region
- build pipelines to refinery
- ship refined oil to oil-fired energy plant
- burn oil to power steam turbines to create electricity
- send electricity out to electrical grid

4 This idea is mentioned in Sim Van Der Ryn and Stuart Cowan, *Ecological Design* (Washington, D.C.: Island Press, 1996).

- transport electricity to end-user (power plant)
- send electricity from power plant to home or business
- put clothes in electric dryer
- turn on dryer; convert electricity to thermal energy to dry clothes

Scenario 2. In scenario 2, you hang your clothes out to dry. This achieves the same result as scenario 1, but with far lighter impacts on the local environment, with almost no expenditure of energy, and without compromising the integrity of place. In southern latitude climates, this is not only an efficient solution, it exposes the wastefulness of Scenario 1. And yet, millions and millions of household in southern climates use the standard drying machine.

Why do so many people use clothes-dryers? Convenience, space availability, and a culture focused on saving time may all be factors. However, if our media wrote about the hidden costs of electric drying, public behavior might change. Equally, if citizens were rewarded financially for hanging clothes out to dry (perhaps with a tax credit), this might also save energy wasted in electric drying. Ironically, some housing complexes in western societies actually don't allow residents to hang clothes outside on clotheslines because it is seen as a form of visual pollution, and thus not acceptable in condominium or townhome communities.

Fast vs. Slow and the Loss of Sense of Place

Across the planet, people have begun noticing that while the "electronic age" has many benefits, there may also be costs to the "fast world" of the twenty-first century. In Italy, the "Slow Food" movement began in 1986; it was started by Carlo Petrini as a way to challenge the globalization of fast food. When McDonald's planned to build a franchise outlet near the Piazza di Spagna in Rome in 1986, Petrini organized a demonstration in which he and his followers brandished bowls of penne as weapons of protest. The demonstration led to the founding of the International Slow Food Movement, which seeks to counter fast food, fast life, non-sustainable food production, and the eroding of local economies. Petrini and others drafted a Slow Food manifesto, which stated that:

> We are enslaved by speed and have all succumbed to the same insidious virus: Fast Life, which disrupts our habits, pervades the privacy of our homes and forces us to eat Fast Foods, which diminish opportunities for conversation, communion, quiet reflection and sensuous pleasure, thus shortchanging the hungers of the soul. In the name of productivity, Fast Life has changed our way of being and threatens our environment and our landscapes. Our defense should begin at the table with Slow Food. Let us rediscover the flavors and savors of regional cooking.[5]

The Slow Food movement emphasizes a set of cooking and eating behaviors that seek to preserve a healthy pace in the enjoyment of food, as well as an emphasis on local and regional products. It speaks of "taste education," the idea of promoting eco-regions, local food products, regional culture and cuisine, organic farming, ethical buying in locally owned marketplaces, and the risks of monoculture. During the 1990s, Slow Food grew and began developing a political agenda—lobbying the EU (European Union) on trade and agricultural policy and working to save endangered foods. The organizations membership grew from 20,000 to 100,000 members in 150 countries around the world. In Latin America, the movement has nearly one hundred local chapters that are

Fig. 3.6 "Slow Food" preparation in a restaurant kitchen

5 www.citymayors.com/environment/slow_cities.html

Fig. 3.7 The pedestrian-scale colonial streets of Cartagena, Colombia offer a poignant example of "slow city" life

seeking to implement its core philosophies.[6]

Following the Slow Food example, the "Slow Cities" movement, or Cittaslow, was also born in Italy. It held its first meeting in Orvieto, Italy in 1999. About thirty towns sent representatives. The Coordinator was Paolo Saturnini, the mayor of Greve, a small town in the Chianti province. The meeting produced a charter that pointed to globalization as a phenomenon that wipes out uniqueness and generates mediocrity. The Charter went on to propose that cities with under 50,000 inhabitants could join the association and use its logo as long as they would agree to follow the fifty-five guiding principles. Among the key principles necessary to be members of the "Slow Cities" movement were: cutting noise pollution, decreasing traffic, increasing green spaces and pedestrian zones, backing farmers who produce local products and the shops and restaurants that market them, and preserving local aesthetic traditions (architecture, historic places, etc.).[7]

As some observers have noted, the Slow Food movement is largely about protecting "territory" or local culture.[8] The idea is to preserve both local ecology (agriculture, vegetation, local soil, water, etc.) and human tradition (cultural practices), both of which allow diverse parts of the planet to generate their own unique foods, styles of consumption, and ways of life. This is seen as an inherently positive thing, and

6 See the Slow Food website at: http://www.slowfood.com/international/4/where-we-are

7 See the Cittaslow charter at: http://www.cittaslow.net/download/DocumentiUfficiali/2009/newcharter%5B1%5D.pdf

8 Heike Mayer and Paul Knox, "Slow Cities: Sustainable Places in a Fast World," *Journal of Urban Affairs*, 28, No. 4, 2006, pp. 321–334.

a pattern that must be preserved. It is thus a movement that connects ecology, human behavior, and regional culture.

Slow Cities simply takes the principles of the Slow Food movement and seeks to apply them to the larger setting of cities themselves. A relatively young organization, the Cittaslow Movement had over fifty cities as members by 2009, mainly from Italy, Germany, Norway, Portugal, Poland, and the U.K. To become members, cities have to initiate a program with Slow City principles, based on sustainable development. They must also make a commitment to heritage preservation and sense of place. For example, candidates for membership in the Slow City group must demonstrate not only support from the traditional local arts and crafts, but also from local industry, whose products lend distinctiveness and identity to a region. They must also commit to conservation, creating green spaces, walkability, making public spaces more appealing, improved public transit, and eco-friendly design.[9]

The Slow Cities and Slow Food movements have evolved because of dissatisfaction with the alternative—the emerging fast world that has come to define human settlement across the planet. Where places were once constructed and molded by local environments and people's values attached to unique places, increasingly in the last few decades the forces of globalization and high technology have significantly altered people–environment relations. The new fast urbanism threatens the planet's future.

Culture, Place Destruction, and Globalization: A Victory for Preservation in the Case of McDonalds vs. the Historic Center of Oaxaca, Mexico

Global fast food and other multinational retail establishments can have the effect of destroying local connection to place and the environment. For example, McDonald's attempted to build a fast food franchise in the heart of historic Oaxaca, Mexico. In 2002, McDonald's Corporation had a seemingly innocent plan—to put the "golden arches" right on

9 Paul L. Knox, "Creating ordinary places: slow cities in a fast world." *Journal of Urban Design*, 10, No. 1, February 2005, 1–11.

Fig. 3.8 The *zocalo*, or *plaza mayor*, in the city center of Oaxaca, Mexico.

the main square of the city of Oaxaca, one of Mexico's sacred colonial centers. At the time, the international fast food giant already owned some 300 franchises in Mexico. It had plans to build one hundred more. By 2011, it had 500 franchises south of the border.[10] But most of the McDonald's restaurants tended to be located in the suburbs, in shopping malls, or near other commercial centers. It was one thing to build a fast food chain on the outskirts of town (such a franchise already existed on the edge of Oaxaca); it was quite another to cantilever a McDonald's into the tightly knit fabric of one of Mexico's most beautiful colonial squares.

But McDonald's Corporation was determined to install fast food right on the nearly 500-year-old *plaza mayor* (main square) of Oaxaca, in a historic district that had been declared a world heritage site by the United Nations' UNESCO organization. Critics cried foul. Gustavo

10 http://www.mcdonalds.com.mx/#NPC:Institutional%231List1

Fig. 3.9 McDonald's restaurant in downtown Mexico City

Esteva, a respected scholar and community leader, told citizens in a public forum that "This is nothing less than a cultural conquest."

A popular protest movement emerged, led by national and local artists, writers, intellectuals, and environmentalists, and backed by worldwide figures—from former First Lady of France Danielle Mitterand to well-known international author of books on Mexican cuisine Diane Kennedy. By 2003, the Oaxaca City Council, under pressure and worldwide scrutiny, voted to deny McDonald's permission to build on the plaza. At stake was not merely the cultural heritage of Oaxaca's downtown historic district, but the symbolic war between global corporatization and the preservation of precious historic resources, which, like the earth itself, once destroyed cannot be recovered.

Chapter 4

Nature and Cities

Introduction

Most of the world's people, including those in less-developed nations, now live in cities. For example, in Latin America, 41 percent of its inhabitants resided in cities in 1950; by 2000, 75 percent of the population called urban areas home.[1] According to the World Bank, Latin America and the Caribbean may be the most urbanized region in the developing world.[2] Like everywhere on the planet, a new kind of city is emerging in Latin America—the "mega city." This new urban form contains populations at a size never seen before (over 10 million in many cases), and spread out across hundreds of square miles, but often with very dense agglomerations. Some examples include: São Paulo (21 million); Mexico City (19.8 million); Buenos Aires (13.1 million); Rio de Janeiro (11.9 million); Lima (9.0 million); Bogotá (8.6 million); and Santiago (5.9 million).[3] With more and

1 See Marcela Cerrutti and Rodolfo Bertoncello, "Urbanization and Internal Migration Patterns in Latin America," Centro de Estudios de Población, Argentina. Paper prepared for Conference on African Migration in Comparative Perspective, Johannesburg, South Africa, 4–7 June, 2003.

2 See World Bank, *Urban Development Brief on Latin America and the Caribbean* at: http://web.worldbank.org/WBSITE/EXTERNAL/COUNTRIES/LACEXT/EXTLACREGTOPURBDEV/0,,contentMDK:20797274~pagePK:34004173~piPK:34003707~theSitePK:841043,00.html.

3 See United Nations, *Population Prospects*, 2007 Revision.

Rank	Megacity	Country	Population
1	Tokyo	Japan	34,500,000
2	Guangzhou	China	25,800,000
3	Jakarta	Indonesia	25,300,000
4	Seoul	South Korea	25,300,000
5	Shanghai	China	25,300,000
6	Mexico City	Mexico	23,200,000
7	Delhi	India	23,000,000
8	New York City	USA	21,500,000
9	***São Paulo***	***Brazil***	***21,100,000***
10	Karachi	Pakistan	21,100,000
11	Mumbai	India	20,800,000
12	Manila[21]	Philippines	20,700,000
13	Los Angeles	USA	17,600,000
14	Osaka	Japan	16,800,000
15	Beijing	China	16,400,000
16	Moscow	Russia	16,200,000
17	Cairo	Egypt	15,700,000
18	Kolkata	India	15,700,000
19	***Buenos Aires***	***Argentina***	***14,300,000***
20	Dhaka	Bangladesh	14,000,000
21	Bangkok	Thailand	13,800,000
22	Tehran	Iran	13,500,000
23	Istanbul	Turkey	13,400,000
24	Lagos	Nigeria	12,700,000
25	***Rio de Janeiro***	***Brazil***	***12,700,000***

Table 4.1 The top 25 most populated megacities in the world. South American megacities are in bold and italics.

more people living in cities, it is critical for the study and practice of sustainability to embrace not only rural and wilderness regions, but also cities.

Cities were originally built to buffer and protect humans from nature, because western cultures (Europe, the US, etc.) traditionally viewed nature as something to be conquered, overcome, and dominated. In this old way of thinking, cities and buildings were viewed as mechanisms for imposing "order" on the wild, unpredictable, and dangerous world of nature. This attitude made it difficult to incorporate an environmentally progressive philosophy into the building and managing of cities. Indeed, much of the nineteenth and early twentieth centuries marked an era when city-building created enormous environmental problems—from air pollution, coal burning, and fireplaces used to warm homes to water contamination from untreated sewage and industrial runoff. Nineteenth century London was often referred to as "the city of dreadful night" due to the combination of soot, fog, and industrial pollution.[4]

The Discovery of Nature and the Environment in Cities

By the early twentieth century, it became clear that cities were becoming severely polluted; leading thinkers recognized the need for better management of nature. In Europe and the US, a number of utopian figures emerged to offer their visions of the importance of rediscovering nature in cities:

- Ebenezer Howard was at the forefront of what became known as the "Garden City" movement. It advocated building smaller towns in a "greenbelt" of preserved space outside the main cities. These "garden cities" would have their own factory work places, a town center, commerce, and open spaces within walking distance, and would be connected by mass transit.
- In Scotland, Patrick Geddes built the "Outlook Tower" in the city of Edinburgh in the late 1800s. This urban observatory sought to study the relationship between city dwellers and the environment, and is considered one of the important turn-of-the-century attempts to better connect city and nature.
- In the US, the tradition of "park planning" was founded by a visionary named Frederick Law Olmstead. Olmstead's plan for Central Park in New York City (1857–73) started a park movement that created vast natural open spaces across cities of North

4 Peter Hall, *Cities of Tomorrow* (London: Blackwell, 1992, pp.13–46.

Fig. 4.1 Portrait of Ebenezer Howard

Fig. 4.2 Portrait of Frank Lloyd Wright

America and inspired those in South America. Olmsted believed that a park should be more than just a green space—it should have the feeling of topography, nature, and diverse landscapes.

- In Chicago, one of the first great city planners, Daniel Burnham, was hired to do a Master Plan for a city that had been partially destroyed by a fire. Burnham's Chicago Plan (1906–09) is one of the first great big city master plans—it paid close attention to the natural landscape, creating ribbons of parks and gardens along Lake Michigan and throughout the city. Burnham coined the oft

Fig. 4.3 Central Park, New York City

quoted phrase "Make no little plans; they have no magic to stir men's blood ... Make big plans; aim high in hope and work."

- Frank Lloyd Wright, an American architect, later echoed Howard's idea in his concept of "Broadacre City," an ideal horizontal city spread across rural landscapes, thus preserving the natural landscape while creating small cities with tall buildings.

Modernism and the Environmental Movement

Despite some of the work of early utopian thinkers, no real environmental movement surfaced until after WW II. In the aftermath of the dropping of two atomic bombs in Japan, the observed devastation following that nuclear holocaust jolted Americans toward creating a modern environmental movement. In 1962, Rachel Carson's book *Silent Spring* spoke of the dangers of pesticides and toxic chemicals for planet earth, especially near cities. Other important books looked at global population growth and resource consumption and pollution, and finally, the idea that limitless growth is not sustainable. A watershed date in the US was 1970, the landmark date for both Earth Day (1970) and the creation of a federal environmental entity—the Environmental Protection Agency (1970).

Fig. 4.4 Freeway movement, one example of "fast urbanism"

The "Environmental Movement" is best understood when viewed in the context of "modernism," the general philosophy that underlay much of the development of cities in the twentieth century. Leading urban and environmental thinkers believed in what they called "machine age cities." They were based on three principles: a) Rationality, b) Science, and c) Order. Modernism put great faith in the ability of technology—trains, cars, roads, elevators, etc.—to solve the problems of building and tearing down nature.

Fig. 4.5 A 747 jet airplane, the technology of a faster world

One way to view the impact of modernity on the environment is to consider the transition from "slow urbanism" to "fast urbanism." Before the industrial revolution, one of the advantages of less technology was that humans existed in a more "slow urbanist" world—people moved around on horseback, on foot, or on river barges, and thus, by the very act of moving more slowly, people were aware of the natural environment. Indeed, pre-industrial cultures also survived because of nature—farmers, fishermen, lumberjacks, and miners worked the land, and thus pre-industrial citizens were attuned to the workings of nature.

The industrial revolution very quickly introduced a new kind of city lifestyle and culture—we might term this "fast urbanism"—a world of trains, automobiles, telephones, factories, mass production, and later on, of cell phones, smart phones, iPads, computers, fax machines, text messaging, and the Internet. This fast-paced modernist society created many luxuries and opened up a dramatically different landscape—freeways, skyscrapers, international trade, and jet travel. But "fast urbanism" also posed new dilemmas for the environment—how to maintain consciousness of nature and ecological preservation amidst the luxuries of the modern age?

Modernist city builders thought they could solve all human problems with technology—whatever problems might exist, they believed engineers, architects, doctors, and scientists could solve them with more investment in science and technology, and more growth. In the fields of architecture, city planning, and environmental design, some of the great "modernists" of our times thought forward-looking, rational, design, and high-tech building solutions could create a new social order. Various schools of thought around the world emerged—the

Bahauas school in Germany and similar movements in France, Switzerland, the US, and Latin America.

The Case of Brasilia

One of the most illustrative examples of the utopian ideals of modernists was the building of the city of Brasilia, the new capital of Brazil, built from 1957–1960. Brasilia was a completely new city, planned from scratch and built in an interior region on a flat plain in the northeast of Brazil. It was a city of highways, planned high rise neighborhoods with grids of "*superquadras*" (super-blocks), a giant artificial lake, and separate "zones" for different activities—a hotel zone, a commercial zone, a zone of banks, residential wings, and so forth. After its completion, critics lamented its numerous flaws:

1. The massive expense of the effort contributed to enormous budget deficits and escalating inflation—which helped spark a military coup in 1964 and subsequently took Brazil in the opposite direction of the showplace of modernity and progress that Brasilia was intended to embody;

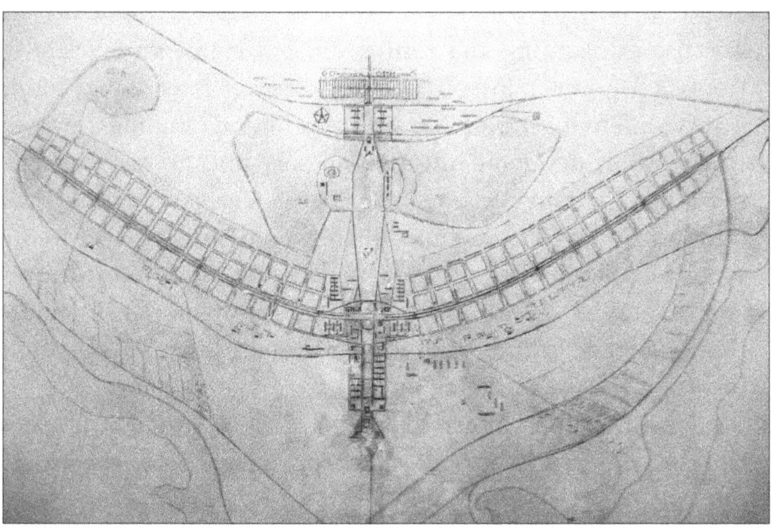

Fig. 4.6 Pilot Plan, Brasilia

Fig. 4.7 High rise public housing: the architecture of modernist town planning

2. Brasilia was designed by planner Lucio Costa as a utopia—its critics argue that its scale and the airplane-shape layout make it difficult to navigate by automobile and not walkable. As critic Robert Hughes wrote about Brasilia, "by design, the pedestrian is an irrelevance."[5] Brasilia was a "machine-age" city, built around the faith in automobiles (and buses) to move people between the different functional zones of the city. This blind faith in the automobile turned out to be environmentally problematic, as the world now knows five centuries later.
3. The use of planning and zoning did not create an efficient metropolis. As writer Joseph Rykwert claims, "Brasilia has become a very different kind of city from the one its founders intended … the whole development seems to have put the very concept of the zoned city into question."[6] Again, the idea of separating people into functional zones assumed there would be a low-cost technology (the automobile) available to move them around from one zone to another. Spreading these zones out in low density, and not thinking out a technological way to connect them by other transit forms, seems short-sighted today.
4. Modernist planners aimed to create a city that would be free of social class differentiation, but slums sprung up before construction

5 Robert Hughes, *The Shock of the New* (New York: McGraw Hill, 1991, p.209)
6 Joseph Rykwert, *The Seduction of Place: The History and Future of Cities* (New York, Oxford University Press, 2000.

was even finished, and by the early twenty-first century, Brasilia had become a highly segregated metropolis—rich and upper

Fig. 4.8 Aerial view of a typical U.S. sprawling suburb

middle class people either lived in medium rise buildings within the original "Pilot Plan" area, near the center, especially around the artificial lake, or in one or two wealthy low rise suburbs. Poor people lived in outer ring "satellite cities" scattered around the Brasilia region.

Many would argue that as a response to the environmental problems of modern urban growth, Brasilia did not succeed, in part because it could not house all of its citizens within the "Pilot Plan"; instead, hundreds of thousands of poor people were scattered in squatter settlements lacking sewage or pollution control. Meanwhile, the dispersed poor must find ways to travel to work back in the planned city of Brasilia, often using buses or old cars, thus contributing to a growing air pollution problem in the region.

Fig. 4.9 An example of sustainable urban design: pedestrian-scale waterfront, Portland, Oregon

The Lessons of "Modernism" for Environmental Management

In Latin America and other less developed areas of the world, government officials embraced modernism as the grand utopian panacea for solving the social and environmental problems associated with growth in the modern era. But there are important critiques of modernism that are fundamental to preserving nature, bio-regions, and quality of life in cities in developing areas like Latin America. They include:

- Buildings without social and ecological planning will fail. Too many modernist projects over-emphasized technology and buildings as the central planning strategy for the future. But without a socio-ecological plan—without creating jobs, affordable housing, sustainable community design, sustainable transit, and environmentally sound manufacturing—futuristic cities will not work. Without planning for diverse neighborhoods within the central urban core, Latin American cities are doomed to become highly polarized, with the upper and middle classes near the center and poor migrants spread out in satellite squatter settlements in the periphery. Brasilia is a good example.
- "Death of the street": high rise buildings alone can be destructive of the sense of place and richness of traditional high density

Fig. 4.10 High tech design for public transit station, Curtiba, Brazil

cities. Modernists built high rise buildings without considering how they fit with the surrounding ecology. Many of the early high-rise city plans lack open space, community facilities, or a plan to promote walking. This makes them highly unsustainable. Density from high rise buildings can be a good thing, if it is linked to larger community design elements.[7]

- Segregation of land uses. Modernist planners argued for new "master plans" and "single use zoning" for cities that separated or segregated different uses of land defined with a rational logic: residential, commercial, industrial, schools, etc. But this had the effect of creating isolated, fragmented, alienating spaces, especially in the form of sprawling suburbs, and it gradually destroyed the sense of place necessary for city neighborhoods to survive. Single use zoning was viewed as a "modern" way to build cities. Urban engineers and government officials thought that by separating people into functional districts, city life would

7 One important element that did make Brasilia's "superquadras" (superblocks) become more neighborhood-oriented was the "entrequadras" (between-blocks) commercial zones laid out between the superblocks. These ended up becoming gathering places for pedestrians from adjacent superblocks. Some of the entrequadras actually have pedestrian-oriented uses, such as cafes, restaurants, or small markets.

be more efficient. In fact, this fragmentation is one of the worst legacies of the modern industrial city. It creates inefficient travel patterns, with people travelling from home to work, work to home, home to shop, shop to home, home to social encounter, etc., thus causing overconsumption of fuel and wasteful excesses in the number of daily intra-city trips.

- Corporatized cities. Modern cities allowed corporations and mega-real estate interests a great deal of power to build and take control of land in the city. Examples include corporate malls, corporate industrial parks, and corporate suburbs. What was missing, and what needs to be done now, is that governments must carefully balance the need for private investment against protecting the public interest (the environment). Too often long-term damage is generated by poorly regulated corporate urban development that is controlled by private business rather than the public.
- Over-dependence on the automobile. The automobile has become the principal machine that defines urban life in the modern era. Even in less developed nations, cars increasingly dominate the lives of workers and city dwellers.
- Wasteful consumption of space and energy—low density suburbanization, shopping mall construction, and strip development are all examples of wasteful patterns of building and subdividing new neighborhoods. The US model of urban industrial growth was built around the separation of residential and work sites. In addition, the size of homes has expanded enormously. For example, in the US the average size of homes in the 1950s was 1100 sq. ft., while the average size of a home in 2000 leaped to 2340 sq. ft. During the same period, family size actually diminished—from 3.7 members per household in the 1950s–1970s to 2.6 members per household in 2000. The use of coal-fired energy production as opposed to natural gas or solar power has depleted coal sources, making energy more expensive, especially for those of limited means.

Best Practices

A number of "best practices" are now being developed to go beyond the utopianism of early modernist thinkers and create truly sustainable cities. Some important criteria for improving the environmental design of cities include:

- "Eco-cities." There is a movement afoot to create cities where nature re-emerges, not merely by planting trees, but by bringing back the streams, hills, canyons, and other natural forms over which the structure of the city was built.
- "Bio-regions." Ecologists are pushing the idea that every city must be better integrated into its true biological region. For example, the principle of "ecological footprint" is now being promoted as a way of measuring the supply of regional resources and what impact a city of a certain size has on its surrounding "bio-region."
- "City of villages." Another way to reconnect cities with nature is to begin to think of creating "villages" within metropolitan areas. These villages would be walkable and have a discrete center with activities necessary to a neighborhood (post office, supermarket, café, copying center, etc.). The idea is to create cities that cut back on wasteful uses of space (big box stores, automobile-scaled suburbs, giant oversized homes) and think in terms of energy efficiency and regional sustainability by creating more centrality and concentration of uses and destinations.

The Example of Curitiba, Brazil

Curitiba, Brazil is a Latin American city that has become well known for being one of the best modern examples of a sustainable city. It achieved this success through a number of strategies:[8]

First, the city took on the idea of integrating traffic management, transportation, and land-use planning. Curitiba created a successful Bus Rapid Transit system and built vital pedestrian zones in the center of the city. Downtown areas were transformed into pedestrian streets,

8 See Local Governments for Sustainability (ICLEI) website "Orienting Urban Planning for Sustainability in Curitiba" www.iclei.org/

Fig. 4.11 One of Latin America's best examples of sustainable urban design: Curitiba, Brazil

including a twenty-four-hour mall with shops, restaurants, and cafes and a street of flowers with gardens tended by street kids.

The all-bus transit network has special bus-only avenues created along well-defined structural axes that were also used to channel the city's growth. The transit system is rapid and cheap and is currently being integrated with the metropolitan region. Its efficiency encourages people to leave their cars at home. Curitiba has the highest public ridership of any Brazilian city (about 2.14 million passengers a day), and it registers the country's lowest rates of air pollution and per capita gas consumption. Creation of an inexpensive "social fare" helped promote equity for poorer residents living on the city's periphery. A standard fare is charged for all trips, meaning shorter rides subsidize longer ones. One fare can take a passenger seventy kilometers.

A visionary open space program was another strategy employed to make Curitiba more sustainable. Curitiba has a network of twenty-eight parks and open spaces. In 1970, there were only about ten square feet of green space per person. Today, there are over 500 square feet for each person. Residents have planted 1.5 million trees along city streets. Builders get tax breaks if their projects include green space. Flood waters diverted into new lakes in parks have solved the problem of dangerous flooding, while also protecting valley floors and riverbanks, acting as a barrier to illegal occupation, and providing aesthetic and recreational value to the thousands of people who use city parks.

Finally, a "green exchange" employment program focuses on social inclusion, benefiting both those in need and the environment. Low-income families living in shantytowns unreachable by truck bring their trash bags to neighborhood centers, where they exchange them for bus tickets and food. This means less city litter and less disease, less garbage dumped in sensitive areas such as rivers, and a better life for the undernourished poor. There's also a program for children where they can exchange recyclable garbage for school supplies, chocolate, toys, and tickets for shows.

Chapter 5

Globalization, Design, and the Environmental Impacts of Tourism Development

One of the problems of global investment is that the decision-makers responsible for development are often not located in the region where that building occurs. This can lead to place destruction; exogenous investors often don't understand the region, lack information, or are not concerned with defending local culture or the environment.

The Homogenization of Design

In the interests of developing worldwide consumption of their products, global corporations seek to homogenize those products—from soft drinks (Coca Cola, Seven Up, etc.) or fast foods like hamburgers (McDonald's, Burger King, etc.) to coffee (Starbucks) and clothing (Gap, Calvin Klein, etc.). This allows global control of consumer behavior through massive, planetary-scale advertising.[1] On the one hand, this ensures consumers that the product will be of the same quality, wherever it is sold. On the other hand, consumers may feel manipulated by global investors trying to reign in their tastes.

This homogenizing principle has spilled over to the built environment. Globally branded hotels, shopping malls, and mega-resorts or

1 This point is emphasized in Leslie Sklair, *The Sociology of Globalization*, (Baltimore: Johns Hopkings, 1990).

Fig. 5.1 The homogenization of design: a bland shopping center façade in Mexico

lifestyle living communities, like fast food outlets, have been spreading across the planet over the last three decades. The diffusion of these global prototypes of design creates landscapes of "sameness" that threaten to eclipse unique culture regions and replace them with bland corporate consumer spaces that look the same in Boston, Buenos Aires, Beijing, or Bombay. This idea of using familiarity and sameness to market a product through the global media has been posed as a "culture-ideology" of consumerism.[2] The culture-ideology of consumerism is turning into a global cultural landscape of consumerism.

Tourism and the Environmental Effects of Globalization

The problem of global manipulation of generic built environments is particularly worrisome for the case of tourism development. Tourism has increasingly become controlled by global actors. The selling of tourism is now a global commodity. It involves the marketing of

2 Ibid.

"place" as a temporary experience for outside visitors. That marketing can have disturbing effects on the local environments of the tourism destinations. Investors in tourism infrastructure may have little knowledge about the place where they invest. As mentioned, they may not be concerned with defending local culture and history against the encroachment of foreign values, including the homogenous design standards of global companies. Those companies are understandably seeking to further the branding of their products. But the insensitivity of global investors to local culture and ecology can lead to the destruction of valuable historic buildings, architecture, even entire neighborhoods, as well as civic culture and local natural resources.

There are at least four ways in which globalization impacts culture and the environment. They include:

Fig. 5.2 Tourism enclave—Acapulco, Mexico

- *Extra-local power*: By its very nature, globalization brings external actors to a local place; these external actors make decisions from afar that have critical and long-lasting local impacts. Tourism is one of many early twenty first century industries increasingly controlled by large, multinational corporations. Global tourist

Fig. 5.3 An aerial view of Cancún, where resorts occupy a strip of beach separated from mainland Mexico.

enterprises—transportation companies (like airlines), travel agencies, advertising firms, and investment groups—increasingly exert a monopoly over the tourism industry. Global corporate entities offer such profitable deals to local operators that they make it almost impossible for them not to participate, even if there are subtle long-term negative environmental impacts. Global investors often fund and construct the built environment tourists occupy—from hotels and resorts to commercial spaces and golf courses. Increasingly, global companies even control some parts of heavily visited sites like historic monuments, through nearby concessions, hotels, restaurants, and commercial spaces.

- *Privatization*: One of the subtle but important impacts of global investment on regions is that not only are the new actors external to the region, they are increasingly private. For example, in the tourism sector, most of the global tourism actors operate within spaces that are increasingly privatized—hotels, gated resorts, exclusive golf or lifestyle vacation zones, private time-share condominium compounds, etc. These spaces have the side-effect of creating tourism enclaves where local residents are actually excluded from their own beaches, mountains, and other natural refuges.

- *Simulation/time space compression*: The changing state of twenty-first century high technology, electronics, computers, and other media is transforming the nature of place. What is real and what is artificial have become blurred. Commercial development has tended toward the artificial, with wildly decorated post-modern malls, festival centers, indoor walkways, and other privatized spaces.[3] All of this leads to what has been termed "time-space compression"—a condition where traditional notions of place are superseded by shifting notions of time and space and of what people call and think of as a place.[4] Scholars argue that places thus become more ephemeral, more short-term and disposable, defined by signs and symbols and simulations. This spills into tourism development, which is increasingly defined by symbols and designs that define the tourist experience, by short-term shifts in the nature of resorts that can change ownership and theme every few years, and by the creation of, and transformation of, resort prototypes. Further, most tourism developments utilize mass marketing of image through the Internet, thus leading to the simulation of the tourist experience, prior to arrival and with a different set of expectations.
- *Enclavism*: An important feature of tourism development often becomes its separation from the reality of the native built environment. It is often the case that tourism zones are physically separated from the host culture. The host cultural setting is sometimes viewed as mildly disturbing to visitors, especially in developing regions like Latin America. Tourism marketing firms also want to protect their clients' consumerism and expend great effort to not allow this to be interrupted. So, for the last three decades in Mexico, for example, tourism resorts have tended to be built in isolated regions, such as Huatulco and Ixtapa, along Mexico's Pacific coast. Tourist districts have ended up as self- contained clusters of western-style hotels, restaurants, and shopping areas. These tourist enclaves tend to be detached "bubbles" of high-priced goods and services, thus enhancing the

3 These are discussed in Michael Sorkin, *Variations on a Theme Park* (New York, Hill and Wang, 1992).

4 David Harvey, *The Condition of Post-Modernity* (Oxford: Blackwell, 1989).

gap between the tourist district and the remainder of the host setting.

One might say that the "Club Med" resort strategy, invented in France, has been extended to the larger tourism industry. Club Med first conceived of the idea of an isolated resort with no telephones or TV in the rooms and all activities largely limited within the resort boundaries. Clients were asked to leave all valuables, including money, at the main desk vaults, since when at the resort they would have everything taken care of. For alcoholic drinks, clients would purchase, in advance, sets of different colored beads that would serve as units of barter within the complex. In this model, there was no need to interact with the outside world.

In the end, the notion of a Club Med-like enclave has spread dramatically across the larger global tourism industry. Enclave design is now the driving force of most tourism development (as it is in the design of residential development in much of the western world—for example, in the form of gated communities). If one observes tourism sites in the twenty-first century, it is not uncommon to see fences, walls, and other images that deliver the message that outsiders are not really welcome.

Fig. 5.4 Aerial view of Acapulco, Mexico

The Commodification of Place

The main goal of the international tourism industry is to enhance marketability and the volume of clients. One way to do that is through the production of landscapes that satisfy the projected needs of users. Studies have shown that tourists prefer comfort, reliability, and pleasure, especially in foreign settings. Urry speaks of "the tourism gaze," a socially constructed tourism landscape that brings to mind a comparison to Foucault's "medical gaze," another systematic, controlled design for an economic interest group.[5]

Thus, tourism developers seek to create a set of homogenous, easily recognized, easily consumed spaces for their client populations. Controlled resort structures with recognizable designs (oceanfront promenades, shopping and restaurant complexes, hotels, fast food outlets, global boutiques) become standard elements of tourism master-planning. The tourism industry, controlled from international command centers in wealthy nations, arguably presents distorted images of developing nations like Mexico in advertising these destinations to clients from wealthier, highly developed nations. Global tourism firms may be less interested in portraying nations as they really are. For example, it is often the case that poverty is minimized or ignored, as are some local customs and practices.

One can ask the question: what incentive can there be for tourism developers to preserve the original landscapes of the places they invest in if the tourism industry earns more profit from homogenized, predictable designs? Even in ecologically sensitive zones (jungles, mountains, river valleys) or culturally preserved spaces (colonial downtowns, indigenous ceremonial towns), the demand for cosmopolitan and consumer-oriented infrastructure by tourists—luxury hotels, swimming pools, fancy restaurants, plush shopping spaces—has the effect of diminishing or overwhelming the original landscapes.

5 John Urry, *The Tourist Gaze* (London: Sage, 1990).

Tourism Development and Environmental Stress in Mexico

Mexico is one of the most important places in the world to examine the impact of tourism on the ecosystem. Mexico became the "star" of international tourism in Latin America in the latter part of the twentieth century, receiving some 40 percent of tourists travelling there. At its peak in the 1990s, twenty million foreigners were annually visiting Mexico, generating over six billion dollars in foreign revenue. During the first decade of the 2000s, tourism continued to draw between twenty and twenty-four million foreign visitors per year, while revenue increased to between ten and fourteen billion dollars annually. These impressive numbers also point to Mexico's big dilemma: how to manage the impact of increasingly large-scale tourism developments and stimulate the economy, but not jeopardize the long-term health of the nation's physical, cultural, and built environments. The Mexican side of the international border is particularly vulnerable to the ill effects of global investment because it is so close to the US and because its economy has been strongly influenced by US capital for more than a century, especially in the tourism and trade sectors. As the saying goes, "pobre Mexico, tan lejos de Dios, tan cerca a los Estados Unidos" ("poor Mexico, so far from God, so close to the United States.")[6]

Fig. 5.5 Map of Baja California

In recent decades, Mexico's tourism development has been accompanied by significant environmental stress. Acapulco was Mexico's original flirtation with the tourism model, the first large scale, palm

6 This quote is often attributed to late nineteenth century Mexican president Porfirio Diaz, though there is some debate as to its true origins.

Fig. 5.6 Upscale hotel in Rosarito Beach, Baja California

tree-lined, tropical resort south of the border. Created after the Cuban revolution in 1959, it grew in two and a half decades from a small town to a giant metropolis of one half million people. At its peak, Acapulco boasted over 12,000 hotel rooms and some 1.5 million yearly foreign visitors. But it also spawned one of Mexico's largest shantytowns. Massive migration toward Acapulco by impoverished rural farmers generated high unemployment, housing shortages, and by the 1970s and 1980s, serious problems of water contamination and air pollution. Acapulco is the first example of catastrophic spontaneous tourism in Mexico.

Armed with the Acapulco experience, the Mexican government attempted several large scale "planned" tourism developments at Ixtapa/Zihuatenejo and Cancun toward the end of the twentieth century. In 1972, the tourism development agency Fondo Nacional de Turismo, FONATUR, in a partnership with the World Bank, systematically designed Ixtapa/Zihuatenejo on the Pacific coast in the state of Guerrero. The idea was to convert the marshes, coconut plantations, and mangrove fields into a modern tourist resort at Ixtapa, while the small town of Zihuatenejo would be converted into a service city via an urban renewal plan. Critics contend that the plan excluded locals, and serious social conflicts arose. As the region grew, low-income laborers arrived and could not afford the escalating cost of housing. An unplanned, haphazard and environmentally destructive corridor of shantytowns emerged along the coast to the east of the city of Zihuatenejo.

Cancun, on a peninsula off the coast of the Yucatan, was also a planned city designed by FONATUR with assistance from the Inter-American Development Bank. It was created using a similar model of a resort paired with a service city. Cancun has been criticized for being destructive both of the physical and the socio-cultural environment. When Cancun was designed, FONATUR's planners did not utilize a comprehensive environmental protection plan; as a result, mangrove wetlands and tropical forests were destroyed. Quarries were dug, creating water-filled holes that became mosquito breeding areas. Wildlife sanctuaries were destroyed and picturesque lagoons filled in for condominium sites. Sensitive coral reefs off the coast were destroyed.

Compromising Baja California

At its peak, eight million US tourists visit northern Baja California each year. Some 40,000 North Americans live permanently on the peninsula, and in one stretch of sixty-five miles along the coast between Tijuana and Ensenada, 25,000 North Americans occupy exclusive housing subdivisions in an enormous wedge of prime beachfront land. Only Guadalajara, Mexico, long a retirement destination for US citizens, has more North Americans.

These statistics do not convey the full extent to which this peninsula was swept into the North American cultural and economic web during the boom years of the 1980s and 1990s.[7] The development of transport infrastructure linking the US with Baja California was one example of growing economic integration. Commercial airline companies opened

7 Of course, after September 11, 2001, the creation of Homeland Security and greater border surveillance slowed down tourism visits to Mexico from the US. The recession beginning in 2008 further hurt the tourism industry, as did the reports of drug cartel murders along the border during the same period.

direct service routes into Baja. Real estate and commercial development projects, funded by US investors, many from the border region, moved south into Baja.

But this growing economic integration had social and environmental costs. The twenty-year boom in northern Baja California generated a number of serious environmental problems. Tourism activities tended to use far greater amounts of water than other land uses and were putting pressure on the state's already limited supply of potable water. Tourism and resort developments were not comprehensively integrated into a regional water/waste management system; therefore, sewage spills into the Pacific Ocean increased. Generally, there is no existing system to account for the frequency and quality of treatment at tourism sites (hotels, resorts, condos, etc.), which reportedly have their own treatment facilities. Further, some of the biggest employers in the region—PEMEX, the Federal Electricity Commission (CFE), and various large fish packing plants in Ensenada—have been cited as sources of toxic waste dumping along the coast.

Crisis in Puerto Nuevo and Puerto Popotla

One example of a place that became globalized and subsequently damaged was the town of Puerto Nuevo, which evolved as a regional tourism town for lobster consumption in the 1970s. It began as a simple place to eat lobsters at low prices in a non-luxurious, family style setting. Like any successful enterprise north of the border, Puerto Nuevo became a victim of its own success. High demand in the 1980s and 1990s caused an absolute boom in the number of establishments and in the scale of construction. The village began with a few small restaurants, all modest in size, all one story. By the late 1990s, there were more than thirty establishments, and many in two and three story buildings. More importantly, worker housing spread to the edges of the town; much of it was built spontaneously, without services or any land regulation. The result was that the growth of the town became precarious, with serious environmental problems including garbage dumping in canyons, cliff-side erosion, sewage spills, and lack of planning. Ironically, during this very same period, Puerto Nuevo's popularity attracted more resort development and new luxury housing, adding density along the coast, in a period when the government did not implement any kind of environmental monitoring.

Puerto Nuevo's commodification and marketing was enhanced by the construction of the Fox Baja Studio for filmmaking and production. This was where the 1997 film *Titanic* was produced. So many visitors came south to view this site that nearby Puerto Nuevo tried to reinvent itself as the marketplace destination for the new Hollywood presence in Baja California.

An example of globalized enclavism and its destruction of place is found at the site of the Fox Baja Studios development in the village of Puerto Popotla, some three miles south of Rosarito. Popotla was once a small fishing village of less than one hundred residents. As mentioned, the Hollywood film company Twentieth Century Fox leased land adjacent to the village to build a major studio for film production in the mid 1990s. The film, *Titanic,* was an enormous global success that had ripple effects on this zone of Baja California.[8]

Residents of the village claim that the government failed to consult them about the impending construction of the film studios, which are so close as to dwarf the town with their massive, ugly metal warehouses and a giant concrete wall surrounding the site, which townspeople have dubbed "the Berlin wall." Security around the site is extremely tight; its entrance on the main highway looks like the entrance to a prison, with high walls and a guardhouse. Here we see "enclavism" in a slightly different form. Some now refer to the Baja Studios complex as a "movie maquiladora."[9]

Fox Baja Studios was a subsidiary of billionaire Rupert Murdoch's global corporate holdings. During construction, the residents of the town claim that underwater explosives, used to grade the beach area and build several giant pools for the *Titanic* filming, caused serious destruction of marine life for kilometers around the site—including kelp beds and magnificent underwater gardens, as well as valuable living creatures like abalone, sea urchins, lobster, and other fish. Further, according to some observers, during filming the film company dumped

8 At least three other Hollywood films used the Fox Baja studios since the "Titanic" production (Fox Baja Studio, 1999)

9 *Maquilas* have long been known to be closed, isolated complexes, sort of walled in "haciendas" of assembly; thus, the term may be quite appropriate.

chlorine into the pool and emptied its tanks in the ocean, allowing the chlorine to seep into the kelp beds and nearby ecology.

The irony, as one explores the landscape of Popotla, is that the very company (Fox) that specializes in making lovely images for movies houses its activities on the edge of the Pacific Ocean in ugly gray warehouses and builds concrete walls and barbed wire fences to separate itself from the town. The scale and style of development, apparently rushed through without any kind of serious environmental impact study or planning analysis, compromised the small village of simple shacks and wooden restaurants for workers.

Chapter 6

Sustainability and Regional Development: The Case of Peru

The Economic Geography of Peru

Peru is the largest Andean nation and third largest country in South America. It is a classic example of a less developed nation exhibiting profound regional inequality, with poverty concentrated in the jungle and sierra (Andes) regions or in poor squatter settlements in the coastal cities, while most of the wealth of the nation continues to be housed along the coast in its largest cities, and in the middle and upper class districts of those cities.

The country is divided into three general ecological regions: costa (coast), sierra (mountains) and selva (jungle)—all of which run the length of the country in a north–south direction. The coast region extends along the western shores of the nation and is characterized by a broad coastal plain that houses the nation's largest cities: Lima, Trujillo, Chiclayo, etc. The coast is also served by a series of river valleys that provide arable land and irrigation to produce food oriented toward the large cities. The coast region is Peru's wealthiest zone; the coastal cities house the vast majority of factories, banks, commerce, offices, and businesses in all of Peru. This imbalance is accentuated by the severe poverty one finds in the other two sub-regions of the nation.

The sierra also runs the length of the nation, and it anchors the parallel ranges of the Andes mountains. Most of the population lives below 9,500 feet, but there are settlements as high as 12,000 feet. The biggest settlements lie in and around the intermontane valleys—the valleys of

Fig. 6.1 A. Santa Maria Beach, coastal Peru, south of Lima. B. Alpamayo peak, Cordillera Blanca range of Andes Mountains, northern Peru C. Peruvian jungle near Bolivian border.

Cajamarca, Mantaro, Huaylas—as well as in the southern end of the country where the Andes widen to the flat plateau or *altiplano*. The large cities in the Andes include Cajamarca, Cuzco, Puno, Arequipa, and Huancayo. The vast majority of residents in the Andes are poor and of Quecha-Indian origin. Their depressed socio-economic state, and the limited economic development of the sierra, contrasts sharply with the urban-industrial life in the coastal zone.

The selva, or tropical forests region, accounts for more than half the entire area of Peru and lies east of the Andes. The jungle region is largely unpopulated except for the high jungle, or *ceja de selva*, and a few major cities along the Amazon and its tributaries—Iquitos, Pucallpa, and Tingo Maria are the main urban centers of the jungle. The jungle's indigenous populations are also poor and perhaps more marginalized than the Andean, Quechua-speaking groups, since they have been quite isolated and cut off from Spanish Peruvian society.

The Roots of Peruvian Poverty and Inequality

As mentioned, Peru's poverty is regionally concentrated in the northern and southern sierra, as well as in the isolated jungle regions that are occupied largely by indigenous Quecha- and Aymara-speaking people or isolated tropical indigenous groups. The dominant causes of inequality include very high concentrations of ownership of land; limited access to capital; unequal access to education, which has often been mainly available to elite coastal Spanish-speaking social groups; and a sociopolitical structure that historically condemned the indigenous rural population to bare subsistence with little chance of mobility.

Fig. 6.2 Indigenous women in a craft guild, Andahuaylillas, Peru

In the post-World War II period, especially since the 1960s, access to education has gradually spread to rural areas. Increased migration to the cities has opened up some new opportunities for people previously trapped in low-wage occupations or rural subsistence farming. Still, poverty continues to abound both in the interior and in the coastal shantytowns.

Social inequalities are rooted in the fundamental structural dualism of Peruvian society, which was historically imposed during the Spanish colonial period. Before the Spanish conquest, the Inca society successfully farmed the Andean valleys, cultivating four times the amount of arable land as it does today. While Inca society was totalitarian and hierarchical, its communal values, ecologically attuned technology, and genius for production and organization allowed indigenous people to prosper from the land before the Spanish arrived.

When the Spanish arrived, they seized the best lands, forced the Indians into slave labor in mining operations and agriculture, forced indigenous people to pay taxes, and otherwise subjugated what was once a noble empire that lived close to the land. Lethal diseases, exploitation, and the psychic shock of conquest led to the demographic and social collapse of the Andean people. This is the legacy of internal colonialism described earlier.

Contemporary Development Strategies

Given the regional inequality that characterizes Peruvian poverty, experts on development have long debated how to pull Peru out of its economic malaise. Generally, there have been two approaches to development. The first approach is labeled "development from above" or "top-down." Development from above approaches grew out of economic theory and the idea that development could be generated by introducing "growth centers" or industrial development poles into poor regions. The idea would be that investing in industrial centers would create jobs and innovation that would "trickle down" to the less advantaged, ultimately transforming poor regions into export-oriented developing zones.

Fig. 6.3 Planned city of Ciudad Guayana, one of Latin America's well-known examples of a "growth pole"

This approach also grew out what was called "growth pole" theory. Growth pole theory treats industry as the basic unit of analysis, one that exists in an abstract economic space. Economic development is viewed as a structural change brought about by investing in "propulsive industries." Propulsive industries are growth-inducing factories built in poor regions. According to the theory, these new complexes of industry would spark innovation that would then diffuse to the surrounding regions and gradually transform them into prosperous zones.

Fig. 6.4 Small village in northeastern Peru

In practice, experts found that development from above often didn't work in poor nations like Peru. Innovations introduced in new towns or cities do not necessarily trickle down to poor people. Too often, growth principles that work in industrialized, wealthy nations don't apply to poor agrarian societies. Thus, even if industrial development centers, or "growth poles," could be introduced into poor Andean regions, elite entrepreneurs frequently controlled profit and paid low wages to indigenous people, who often drift away from these factories. Marginal regions end up simply supplying commodities and cheap labor, while the economic power still remains in the big cities on the coast.

An alternative approach has been called "development from below". This "bottom-up" model emphasized growth tied to local ecology and culture. Development from below recognized that a successful economic development strategy was needed to empower poor, marginalized sub-regions to overcome the centralization of political and economic power that dominated countries like Peru. It was an early version of what later grew into the sustainable development movement, which advocated grassroots development tailored to the unique history, culture, and ecology of specific sub-regions of poor nations like Peru. It also called for the use of appropriate technologies and ecologically sustainable practices that preserved the natural environment.

Fig. 6.5 Periodic market or "feria," Písac, Peru

Integrated Regional Development in the Central Andes: A Case Study

The central Andes still displays the inequality patterns described earlier. These include:

- Impoverished indigenous people living in medium-density towns and cities spread across the mountains
- Subsistence agriculture
- Widespread poverty
- High levels of illiteracy
- Poor nutrition and health problems
- Little knowledge of how to market agricultural produce

To address these inequities, a sustainable development approach must address the specific needs of rural settlements, from social services and supplies to roads, warehouses, markets, and other facilities that enhance local farmers' ability to not only grow food, but to store it and market it to the large cities along the coast. Rather than bringing in outside industry, which does not fit the culture or ecology of the region, an alternative approach would be to create localized strategies that focus on two kinds of investments: social infrastructure (water, electricity, schools) and economic infrastructure (roads, irrigation systems, warehouses, markets, slaughterhouses).

The Role of Cultural Institutions

One critical discovery made by development planners and researchers in Andean Peru was the importance of "key market towns" or "*ferias.*" *Ferias*—also called periodic markets—are a well known institution that has been embedded in Andean culture for centuries. Long known as weekend or once a week markets that brought buyers and traders together from distant villages and towns, key market towns had a number of important characteristics that made them strategic locales for local economic development. These characteristics include:

- *Linking buyers and sellers.* One of the amazing qualities of *ferias* is that they are well entrenched in the ecology and social life of rural farmers. People know where the periodic markets are located and how to get there. They are an evolved and polished system of trading and marketing tuned to the regional geography of the Andes.

Fig. 6.6 Vendors in periodic market, Chinchero, province of Urubamba, Peru

- *Providing access to high order goods.* Isolated, rural peasants would otherwise not have access. Because people travel many miles by foot, pack animal, truck, or bus, they bring a variety of goods for trading to towns that increasingly sell things they would otherwise not be able to buy. Again, the *feria* system is a very rational bartering program.
- *Performing specialized functions* (slaughtering, for example). In the more important periodic markets, ancillary services have evolved. For example, in towns that sell live animals, locally skilled butchers provide slaughtering and butchering services to potential buyers. Meat is cut from animal carcasses; hides are skinned; both are offered to buyers.
- *Offering a venue for vital social activities* (feasts, courting, or weddings). The market towns also have a social function: they provide a meeting place for young people to court and get married. Without the *ferias*, many indigenous young people would be isolated on farms and in very small villages. The periodic market system creates a social networking arrangement that transcends the small village.
- *Creating sites for religious services* (visit to regional churches). Catholic churches locate in the most important periodic market towns, thus making specialized religious services (baptism, holidays, etc.) available to poor Indians if they choose it.
- *Enhancing political participation.* The *ferias* serve as a center, allowing people to vote and interact with government officials. For example, indigenous people learn about the national tax system through information they receive in the periodic markets.

Periodic Markets and Sustainable Development

What we learn in Peru's Andean highland is that an ancient tradition—the *feria*, or weekly periodic market—can provide a key institution that serves as a catalyst for locally-based sustainable development. Several lessons are provided by this cultural tradition.

First, the identification of investment locales should be not only tied to the population of a town or urban center, but to its cultural relevance to surrounding residents. The weekend market (*feria)* towns were not always the most populated places, but, over time, they carved a regional role in centralizing the barter system. As gathering places within the geographically dispersed populations of indigenous farmers across the Andean mountain region, periodic markets become critical to any strategy that seeks to reach more people, deliver services to them, and get them more involved in their own regional economies.

Second, a successful sustainable development strategy in the Peruvian Andes depends on local knowledge of patterns of movement, and on knowing which towns need certain types of investments—social infrastructure or economic infrastructure. By understanding the nature of bio-regions (terrain, soils, rainfall, cultivation patterns), local production behaviors, and the condition of existing infrastructure (roads, water, electricity, schools, etc.), a sustainable development plan can be tied closely to the needs of poor farmers.

Finally, key market towns are essential to service delivery and teaching farmers to market their own products. Because periodic markets have such a strong centripetal pull, they attract large populations and become strategic locales for delivering social and other services. They also become economic development poles—marketing centers for regional agricultural output, local crafts, and other locally produced goods and services.

Chapter 7

Sustainable Cities

The problem of harmonizing nature with human settlement ultimately must be confronted in cities because such a large proportion of people on this planet now live in urban areas. In 1950, there were eighty-three cities with populations exceeding one million; by 2007, this number had risen to 468. Demographic projections suggest that the world's urban population of 3.2 billion will grow to about five billion by 2030; by then, three out of five people will live in cities.[1] Between 1950 and 2000, Latin America's population increased from 160 million to about one-half billion. During that same period, the urban population grew to nearly 400 million, with some three-quarters of its citizens living in cities. The vast majority of "mega cities" (over ten million population) in the world lie in developing nations—Jakarta, Bombay, Delhi, Manila, Cairo, etc.

Several booming megacities are located in Latin America—Mexico City, Sao Paulo, Buenos Aires, and Rio de Janeiro. The environmental and quality of life problems faced in these and other large cities stem largely from uncontrolled growth. The colonial period placed an emphasis on exploitation of natural resources and administering growth from capital cities. During the industrial revolution, growth began to accelerate, yet there were no professional disciplines—environmental management, city planning, or land regulation—to begin to handle

1 United Nations, Department of Economic and Social Affairs, Population Division (2006). World Urbanization Prospects: The 2005 Revision. Working Paper.

this growth. By the twentieth century, a massive population shift from rural areas brought millions of people into large cities that were not prepared to absorb them. Urbanized regions have proven to be ill-equipped to manage the scale and form that growth has taken.

Fig. 7.1 View of Rocinha favela, Rio de Janeiro, Brazil, one of the largest squatter settlements in Latin America

The Squatter Settlement Problem

The combination of rural–urban migration and high birth rates in Latin America during the second half of the twentieth century permanently transformed the nature of the Latin American city—it saw the growth of giant rings of poor squatter cities outside the original urban core zone. The "squatter settlements" were spontaneous land invasions of poor migrants living on marginal land, initially with scrap materials (tar paper, wood, woven mats, etc.) and later with whatever could be purchased (brick, concrete block, stone). Squatter communities represent a "self help" style of survival among millions of rural poor who lack the financial resources to either rent or purchase conventional housing. Instead, they choose to risk squatting illegally on undesirable land, with the hope that the government will allow them to stay.

The steps taken in squatter community formation include invasion of land, incremental construction of homes as unstable sources of income allow, and the securing of services (piped water, electricity, paved roads, schools, etc.) either illegally (by pirating electricity or water through unmetered, makeshift connections) or through negotiation with local, state, and often federal governments. One of the biggest problems facing governments is lack of regulation of squatter settlement growth and planning. In many cases, spontaneous urbanization occurs in ecologically sensitive zones—steep sloping mountainsides, flood-prone river plains, on the slopes of volcanoes, and in swamps. These are environmentally sensitive locations where local government planning offices should either prohibit or strictly regulate growth to protect residents; however, given the absence of resources to provide alternative shelter for the poor, governments have often chosen to look the other way. Equally, many shantytowns are located near toxic waste dumps, garbage incineration sites, heavy industry, or airports. Earthquakes, urban fires, and gas explosions often impact squatter neighborhoods. This exposes residents to unusually high levels of air pollution, industrial sludge and waste, noise, dust, and other unhealthy

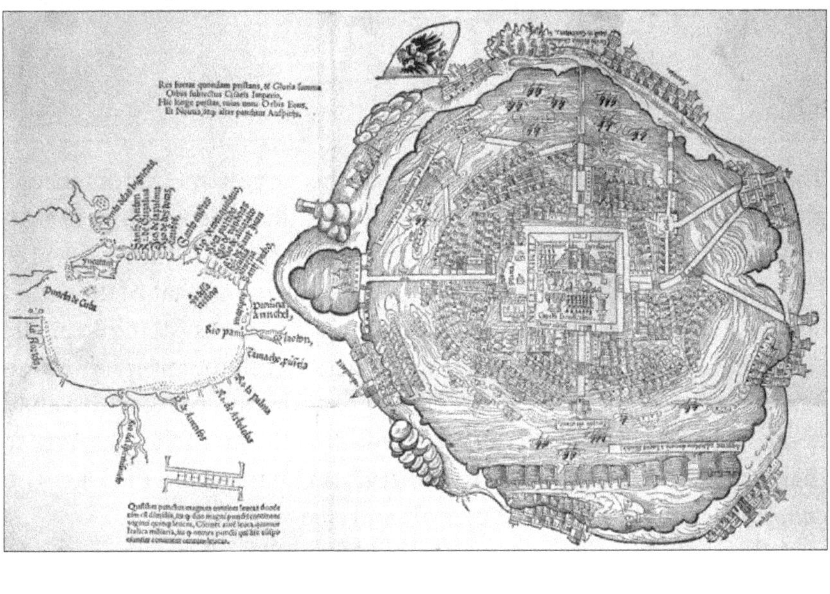

Fig. 7.2 Hernán Cortés' map of Tenochtitlan, 1524

Fig. 7.3 Aerial view of Mexico City metropolis, with smog visible on horizon

contaminants. Children are especially vulnerable to toxic air and dust, which can seriously damage their more delicate lungs. The ecological problems that plague squatter communities, combined with uncertain legal title to their properties, make these vast satellite slum zones spaces of uncertainty and potential political instability in the future.

The Case of Mexico City

The Valley of Mexico encloses one of the largest metropolitan agglomerations in the world. Some twenty million people live in and around this valley and its primary center, Mexico City. During the pre-Hispanic period, the basin of Mexico City was the capital of the Aztec empire and prospered. The Aztecs built an intricate system of canals, causeways, dykes, and aqueducts to harness the abundant water in the basin. To grow food, they built "*chinampas*," or land plots, which they floated and drew water and nutrients for directly out of the lake, rather than trying to pipe lake water to the dry land beds around the lake. The *chinampas* were built from canal silt and marsh weeds.

When the Spanish conquered Mexico City, they immediately destroyed most of the ecological infrastructure the Aztecs had put in place. The Spanish wanted open roads for their armies and space to build neighborhoods for colonists. They demolished causeways and aqueducts and filled in canals. By the early 1800s, local water was running out, so the Spanish (and later Mexican) government made one of the biggest blunders in urban environmental history—they began drilling into the aquifer beneath the city to find water. During the late nineteenth century administration of Porfirio Diaz, the Mexican president hired engineers to build a giant canal, draining water out of Lake Texcoco for Mexico City residents.[2] In the twentieth century, the government continued to draw water from aquifers deep within the basin beneath the city. The drilling of the lakebed on which millions now live has permanently altered the subsoil foundation of one of the largest cities in the world. This has led to increasing instability, as evidenced by the events of the 1985 earthquake. This raises the question—can a mega-city of this size sustain itself on the silt-like, shaky plates of a former lake bed?

Fig. 7.4 Smog hovers over Mexico City mega-region

Urban Growth: Three Critical Environmental Problems

Mexico City grew from 3.1 million in 1950 to 13.9 million in 1980, and by 2010, it had over 19.5 million people, making it the ninth largest metropolitan area in the world. This growth has occurred, in

2 An excellent source is Jonathan Kandell, *La Capital, The Biography of Mexico City* (New York: Random House, 1988).

Fig. 7.5 Poor squatter neighborhood in northern Mexico City, Colonia Atlampa

part, because the national political system favors the concentration of industry, banks, commerce, and political power in the national capital, Mexico City. For many decades, national investment has favored energy, highway, water and drainage facilities, and other infrastructure in the Mexico City basin, thus encouraging more centralization. Urban experts refer to this phenomenon as "primacy"—when one city has an overwhelmingly high concentration of population and economic activities of an entire nation.

The most important manufacturing activities in Mexico City include: textiles; furniture and repairs; publishing activities; production of rubber, plastic, and metal goods; and the assembly and repair of electrical goods. For most of the twentieth century, this production was for national and local markets, but by the late twentieth century, it became oriented towards global markets, especially after Mexico signed on to the North American Free Trade Agreement in 1992.

The expansion of Mexico City and the surrounding valley has had serious impacts on the ecosystem in three areas:

Air pollution

The Mexico City basin is 6800 feet high and enclosed by mountains. There is less oxygen at this altitude. Auto pollution, unregulated industry, household stove usage, and massive dust created by draining the soils of what was once a lake are the main sources of air pollution. There are some 3.5 million private autos in the city—and daily traffic congestion makes pollution worse. There are also 35,000 factories in

Fig. 7.6 View of Rio de Janeiro and Guanabara Bay, from Corcovado Mountain

the city, generating noxious gases. Intensive sunlight transforms these gases into a thick smog, which chokes residents. When mixed with the altitude and low oxygen levels, polluted air has become elevated to a national public health crisis. Emphysema, pneumonia, bronchitis, asthma, and cardio-vascular illness have all been scientifically linked with toxic air in the Mexico City basin.

Water scarcity

Mexico City's water supply lies more than sixty miles from the city and 3000 feet lower. It costs nearly $100 million a year to produce the energy needed to transport distant water into the city. Meanwhile, only 75 percent of all households have in-house water, while nearly one third of all homes lack adequate sewage facilities, mainly in the poor *colonias* on the outskirts of the region. Water is extremely expensive because it is so far away; ironically, by continuing to extract it, the city is also causing further ecological damage. Draining the aquifer lake creates more dust, and thus more air pollution. Further, as mentioned earlier, the city's geological base has become more unstable (certain parts of the city are literally sinking, for example), in part due to constant drilling into the soils beneath the city. This instability drew worldwide attention during the 1985 earthquake, when sandy and unstable soils caused neighborhoods to shake violently, while tall buildings subsequently toppled. The most damaged zones in the city were those in the bed of the historic Lake Texcoco, where the prevailing silt and volcanic clay sediments amplified the shaking. Over 10,000 people died, 30,000 were injured, and 100,000 were left homeless. Major hospitals, hotels, and residential tower buildings were destroyed. Two decades later, the city is still rebuilding and reinforcing existing structures.

Housing shortages and human settlement

The housing shortage in Mexico City amounts to nearly three million families without homes, according to the government. The high cost of land forces poor people to seek alternative shelter in "irregular settlements" or unregulated *colonias* on the periphery of the metropolitan region. These often occur on plots of land illegally subdivided by landlords, real estate companies, or even "*ejido*" (land communes) land owners. The resulting giant peripheral settlements are poorly serviced and mostly unregulated. The worst *colonias* are located in high-risk, heavily polluted areas in the north and east, near industrial zones, the airport, and other less attractive neighborhood amenities. Residents generally have questionable property titles. Irregular settlements constitute roughly one-half of the Mexico City urbanized area and house more than 60 percent of the population.

Although federal government planning strategies strive for the decentralization of growth away from Mexico City, tax subsidies and other public policy decisions often make the nation's capital a more attractive place to live and work than smaller cities in other regions. Furthermore, Mexicans who desire to remain in Mexico City are influenced by numerous social, political, educational, and cultural factors, and they often equate living in Mexico City with the image of personal success.

The Case of Rio de Janeiro

Rio de Janeiro, with a metropolitan region population of nearly twelve million, is the second largest urban area in Brazil, and the fourth largest in Latin America (after Sao Paulo, Mexico City, and Buenos Aires). Rio is a region with dramatic mountains, bays, rivers, islands, and coastal settings. But its rapid growth is also posing serious challenges for sustaining its geographic beauty.

The city's growth and development, as early as the colonial period, was a process of "conquering nature"—overcoming the constraints of its physical geography—the lowlands, mangroves, swamps, and hills. This was consistent with the western (European) idea that progress is measured by human conquest and control of nature. In Rio, this unfortunately involved gradual destruction of the very environmental beauty that attracted people to it in the first place and that continues to be a huge part of the city's identity, and thus of the economic base that currently sustains it. Thus, for Rio, "natural beauty" is not merely a side benefit of living there, it is a fundamental public good that must be preserved for the city to survive in the global economy. Rio is thus an excellent example of the overlapping agendas of environmental and economic preservation.

Green Landscape Traditions

Rio de Janeiro has a history of cultural practices and traditions that make the "greening" of the city more possible. Older communities exhibit qualities of good urban design and sustainability. Many neighborhoods interface comfortably with the natural environment. The high-density high rise condominiums, office buildings, and stores along bus-lined commercial boulevards interweave with the largest urban rain forest on the planet. The Austrian writer Stefan Zweig, who visited Rio in the early 1940s, once wrote this about the city:

Fig. 7.7 Green landscape tradition: Pargue del Este, Caracas, Venezuela, designed by Roberto Burle Marx

Fig. 7.8 The diverse ecosystem of Rio de Janeiro includes the rainforest, mountains, bays, and the ocean.

> The city and the sea and the green and the mountains, all these flow into each other like music.[3]

Rio's tradition of place-makers carving parks, gardens, patios, promenades, plazas, and living spaces into the green landscape can be traced to landscape designer Roberto Burle Marx. Marx formalized built landscape connection to Brazil's unique ecosystem, drawing attention to the flora within the vocabulary of modern design. His work jolted his twentieth century peers to better utilize public spaces in Rio and other cities. In the words of one writer, "By organizing native plants in accordance with the aesthetic principles of the artistic vanguard, especially Cubism and abstractionism, he created a new and modern grammar for international landscape design."[4]

3 Stephan Zweig, *Brazil: Land of the Future* (Stockholm: Bermann Fisher, 1941), p. 173.

4 Larry Rohter, "A New Look at the Multitalented Man Who Made Tropical Landscaping an Art." New York Times, January 20, 2009. http://www.nytimes.com/2009/01/21/arts/design/21burl.html

Fig. 7.9 Parque Lage epitomizes organic design traditions in Rio.

Many believe that Burle Marx single handedly transformed Rio, both by his incredible portfolio of major design works in the city and by exciting others to rethink the links between nature and the built city. For example, Rio's largest park, the bayside Aterro do Flamengo, built on reclaimed seafront just southwest of downtown, was Burle Marx's accomplishment. He also was responsible for colorful abstract stone mosaics along the Copacabana beachfront promenade, and for crafting or inspiring numerous other parks, gardens, museum patios, and other spaces around the city, not to mention the proliferation of small nurseries that sell plants and the push to preserve open spaces and jungle in and around the city.

Of course, nothing contributes more to the injection of serenity into a booming city than the presence of water, and like other tropical waterfront cities, Rio is a city of water—bays, rivers, lakes, marshes, lagoons, streams, canals, and of course, the great Atlantic ocean. Water is as fundamental as the rich green jungle landscape.

Fig. 7.10 Despite its physical beauty, Guanabara Bay in Rio de Janeiro faces serious pollution because it lies within one of the largest industrial zones in Brazil

Fig. 7.11 Botanical Gardens in Rio's Jardim Botanico neighborhood.

Aside from the jungle and the sea, and beyond formal landscape design, there is also a local cultural affinity for what one might call "tropical organic infrastructure", which includes parks, plazas, street markets, flower stands, kiosks, and food vendors. The life of the street is inherently tied to nature, both in the form of literally selling products derived from local ecology (fruits, vegetables, herbs, fish, etc.), and in handicrafts that also derive from nature (necklaces, jewelry, musical instruments, etc.). The well-known "Feira Hippie" (Hippie Market) in Ipanema is a place where craftspeople who live in nearby favelas sell organically made products.

These cultural practices in open space point to an organic street life. All social classes in Rio enjoy various forms of engaging with nature through public spaces. The weekly street markets and cultural niche markets are one such example. The Sunday ritual of walking along the promenade in Ipanema and Copacabana, where the streets are closed off and reserved for pedestrians, is another.

Land Use and Sustainability

Land and its uses are fundamental to Rio's uniqueness, but land development is also critical to the present environmental crisis. During the early period of colonial settlement, Portuguese settlers began encroaching on sensitive environmental areas, such as the steep slopes of hills leading to the mountains that ring the city, and the lagoons and the water-clogged flood plains along the coast. By the early twentieth century, these urbanizing lands began to severely flood during the rainy seasons, while the hillsides suffered from erosion and landslides.

Most of the worst landslides occurred in hillside segments that had been cut and then filled by settlers, thus altering the contour of hill slope geography and the natural ecology of drainage toward the sea. As trees and natural ground cover were removed by building and construction, the land surface lost even more water retention capability. Naturally, this made the flooding problem worse. Many of the worst land use violations occurred in areas where poor people moved, often on subdivisions built illegally by unscrupulous developers and landlords. Rio's poor reside in squatter settlements called "favelas." After the 1940s, favela residents migrated into Rio and began occupying the

Fig. 7.12 Urban environmental planning in Brazil must include the nation's favelas

steep-sloping hills around downtown, to the north, and in the mountainous areas of the south zone—Gavea, Tijuca, and Jacarepagua. They also moved into tidal estuary zones and rivers near the coast and were frequently wiped out by mudslides and floods.

Meanwhile, the metropolitan region also began to suffer deforestation. Severe erosion, degradation of water sources, and urbanization have wiped out 40 percent of the vegetation in the Atlantic Rain Forest (the *Mata Atlantica*) of the Rio region. This further diminishes the water retention capability of soils and increases the volume of downstream river flows toward the coast. When beach areas are urbanized, development causes shifts in the flow of silt down-river to replenish beach sands. Money spent to address beach erosion in wealthy neighborhoods in Rio's "South Zone" often depletes the public resources available to help other areas of the city. For example, in the Leblon neighborhood, near Ipanema, roads have been built to provide better access to the beaches and tourist facilities. But some of these roads interfere with beach ecology, creating more erosion and requiring more expenditures to fix those new problems.

Water Drainage and Air Quality

In the Rio metropolitan region, it is critical to allow for the natural flow of water. The gradual filling of the mangrove swamps and lagoons and the removal of vegetation along the slopes of the hills around Rio have severely interrupted drainage. Rainwater has fewer outlets to the ocean and moves faster and at greater volumes. The result is flooding and landslides, especially in the low-lying floodplains near the rivers, river beds, and lagoons. For example, in São Conrado, a wealthy South Zone neighborhood, five rivers were filled in to build the Morro Dois Irmãos tunnel linking the area to Ipanema and Leblon. This shocked local hydrological systems, forcing the Carioca (Rio) government to build an artificial channel for the water, at great expense. Meanwhile, in the hills above São Conrado, one of the largest favelas in Brazil grew—it is called Rocinha. The growth pattern of Rocinha further disrupts the flow of water and increases the problem of flooding.

Toxic wastes and pathogenic organisms are polluting the watershed of Rio, from its two great bays to the coastal seas around the region. Point sources of this contamination include poorly treated domestic sewage and industrial effluent from chemicals, petrochemicals, oil refineries, and iron and steel production, all of which flow downhill into the region's watershed. Most collected sewage in Rio is untreated and simply dumped back into rivers and the ocean.[5] Another problem is clandestine household sewage lines, which are illegally connected to the city's rainwater drainage system, thus mixing waste with rainwater. Non-point pollution sources include uncontrolled domestic waste, agriculture runoff, and storm water runoff.

There are many forms of industrial pollution that impact the bays and the ocean. Oil, heavy metals, and industrial waste are the three biggest industrial contaminants entering the two large bays: Sepetiba Bay, which is thirty-seven miles south of the city, and Guanabara Bay, which lies in the heart of the urban area, adjacent to downtown. Sepetiba Bay is an important fishing region for the state of Rio. But it is seriously contaminated by metals like zinc, lead, chrome, and copper from nearby milling and smelting plants. The mangroves in Sepetiba

5 One exception is the Ipanema outfall, where treated sewage is dumped about two miles offshore.

Bay are so thick that they tend to cause metals in the water to accumulate, thus making the problem worse.

Guanabara Bay has 10,000 factories, ten oil terminals, twelve shipyards, two oil refineries, and is the second largest industrial zone in Brazil. Four hundred sixty-five tons of organic discharge are produced daily, of which only sixty-eight tons (15 percent) are treated. The rivers draining the Guanabara Bay watershed—Iguacu and Estrela—are heavily polluted, and they produce a strong hydrogen-sulfide smell that affects many poor favelas in the north zone and near the airport.

Air quality has become a more serious concern in Rio over the last decade. The sources of air pollution include industry, auto emissions from gasoline, and formaldehyde and acetaldehyde emissions from ethanol fuel emissions. Garbage incineration is also a problem in high-density neighborhoods.

Toward a More Sustainable Future

The Rio metropolis completed an Environmental Master Plan in the 1990s. Among the goals of environmental and sustainable policy in Rio in the future are:[6]

- More rational uses of land. The city government needs to control land speculation and illegal sales and occupation of land. It also needs to control land filling and waste disposal, so existing residents are not endangered either by pollution or by flooding.
- Macro-drainage and canalization. Rio needs an environmental policy to systemize the drainage hydrology of the region—canals need to be created where water flow has been blocked, and settlements on flood plains need to be protected.
- Cleaning up Sepetiba Bay. Rio's most important fishing zone must be carefully protected from metal pollution. Nearby smelters must be forced to use filters and better management of waste and emissions.

6 These are discussed in, Kreimer, Alcira; Lobo, Thereza; Menezes, Braz; Munasinghe, Mohan; Parker, Ronald eds. *Towards a sustainable urban environment : the Rio de Janeiro study, Volume 1* World Bank Discussion Paper. no.WDP 195, 1993.

- Deforestation. A comprehensive land use policy must regulate development on steep sloping hills. A "Hillside Ordinance" should be passed, in which any land with a significant slope is subject to special regulation of development to mitigate slides, flooding, and disruption of existing ecology.
- Illegal household discharges. Steps must be taken to regulate illegal connections of household sewage discharges to the rainwater drainage system.
- Industrial waste. The municipal government needs to regulate and impose fines on factories that dump toxic pollutants into the region's twenty-five rivers.
- Favela management. The city needs to regularize land ownership in favelas and build relations with favela residents to control pollution, illegal sewage dumping, and other ecological threats to residents and to the city. The "Favela Bairro" program has created incentives to improve the quality of life and design of infrastructure in some favelas. This project needs to be infused with a more sustainable and environmental policy orientation.

Chapter 8

Sustainable Policy-Making

Introduction

How do we put the ideas of sustainable development into practice? Two important theoretical perspectives should guide our future thinking: "land ethics" and "steady state economy."

The phrase "land ethics" refers to the relations between individuals and land; it's derived from the general principle of ethics, or "relations between individuals and society."[1] If we have an ethical obligation to master our connection to each other, the argument goes, so too should we have a moral obligation to improve our connection to our surrounding environment (land). In western culture, there is a tendency to value private land ownership over the commons or "public interest" (air, water, flora, fauna, etc.). Private property was a cornerstone of the nineteenth century industrial revolution and the

Fig. 8.1 Image of Aldo Leopold, who coined the idea of "land ethics"

1 See Aldo Leopold, "The Land Ethic," A Sand County Almanac, 1949. In Stephen W. Wheeler and Timothy Beatley, eds. *The Sustainable Urban Development Reader* (New York: Routledge, 2004), pp. 20–29.

economic boom of the twentieth century. In the new millennium we are beginning to realize the cost of ignoring the collective value of nature and natural resources—forests, trees, animals, etc. We therefore need to develop a "land conscience," which would expand the notion of "community" to include the natural ecosystem—soils, water, plants, air, and so forth.

We have seen that the traditional western economic model is based on the assumption that growth is always good. It is also assumed that growth entails cyclical changes—periods of rapid expansion followed by eras of retrenchment. Some see this as a natural process of success and failure in business, a kind of "creative destruction," as one scholar termed it.[2] Yet this roller coaster vision of our economy too often has left the marginalized populations even worse off than before. Further, it underestimates the importance of nature and the built environment, since the cyclical changes often involve tearing down or rebuilding humanly constructed landscapes. It diminishes the value of good urban design, architecture, and town planning. From an environmental design perspective, cycles of "creative destruction" are not sustainable. An alternate approach, termed "steady state," views a world in which governments would regulate the consumption of resources based on their availability. Since the ecosystem is finite, it makes sense to have a human economy that evolves toward a steady state condition, as opposed to one of limitless growth.[3]

The Perspective of "Natural Capitalism"

Natural capitalism offers an important paradigm shift in how we view the environment. Conventional capitalism is based on several ideas: the need for bigger, more efficient factories; the highest and best use of land to maximize production and profit; and the overall measure of human well-being expressed by the nation's output or gross national product (GNP). In this scenario, the natural environment has

2 Joseph Schumpeter, *Capitalism, Socialism and Democracy* (New York: Allen Unwin, 1976).

3 Herman Daly, "The Steady State Economy," from Herman Daly, *Towards a Steady State Economy* (San Francisco: W.H.Freeman, 1973).

Fig. 8.2 Natural capital has enormous economic value: waterfall in Tapanti National Park, Costa Rica

significance only in so far as it contributes to economic growth and improvements in standards of living.

Natural capitalism, on the other hand, argues that we can no longer operate our economy under the assumption that natural resources are only valuable when they are exploited for profit and growth. On the contrary, the natural environment must be valued for its own sake, and for its contribution to human survival and well-being. In natural capitalism the environment must be *the* central factor that drives our economy. Natural capitalism is based on several innovative ideas:

- The environment is not a minor "factor of production," but is the main envelope containing, provisioning, and sustaining the entire economy.
- Economic development is mediated by natural capital and must find a balance between the preservation of natural resources and the goals of growth and development.
- The loss of natural capital is caused by wasteful patterns of consumption, driven by unchecked population growth and badly designed (i.e. polluting) businesses.

Fig. 8.3 Velcro exemplifies the use of bio-mimicry for modern purposes; its design (right) has copied the small flexible hooks of this burr (left), which allow it to attach to soft or fibrous surfaces.

- "Resource productivity" should be measured in terms of four kinds of capital: human, industrial, financial, and natural.

Four Strategies of Natural Capitalism

The four main strategies of natural capitalism are:

- *Radical resource productivity:* involves obtaining the same utility (or more) from a product or process while using less material and energy. Radical resource productivity suggests that technology and creative design can make a product far more environmentally friendly than we currently imagine.
- *Bio-mimicry (green industry):* is the idea that industry should be more attuned to the natural environment. Manufacturing can reduce or eliminate waste; industries can cut back on wasteful activities. Greater industrial awareness can significantly reduce pollution, acid rain, and greenhouse gases. This will allow for the protection of "public goods": water, clean air, forests, and veins of minerals.
- *Service and flow:* In traditional industry, the connection is between producer and consumer (between goods and buyer). Service and flow seeks to create future industries where the "service–flow" connection moves away from a consumer culture that

targets only the acquisition of the good, toward one that emphasizes its quality, performance, and impact on the environment. Consider, for example, a washing machine: instead of a consumer purchasing the machine, the consumer would purchase "use" of a machine, while the company would own it and be responsible for both the machine and its waste byproducts.

- *Increased energy efficiency*: involves producing and delivering goods and services to minimize the use of energy and the production of non-biodegradable by-products, thus leading to further savings in natural capital.[4]

Producing a "hyper car," or hybrid-electric car with ultra light design, low drag, and electric propulsion, would be an example of using the strategies of natural capitalism. An entrepreneur who invests in this project would earn profit but would be investing in a product that fulfills the four principles of natural capitalism. First, the car's design is aimed at the most advanced resource productivity, that is, delivering the same functions of a traditional car, but with less waste and more environmentally friendly materials. Second, by engaging in "bio-mimicry" and practices to "close the loop" between production and waste disposal, the car is designed to minimize toxic by-products, thus enhancing its longevity. Third, the car would use the "service and flow" model: it would be leased by consumers as a service, rather than purchased as a product. And finally, by being more energy efficient and less polluting, a hyper-car would represent a huge savings in natural capital.

Green Manufacturing: Cradle to Cradle

Traditional manufacturing has always operated with a "cradle to grave" approach. Products are designed to become obsolete. The driving force in product design has been profit, with little thought about longevity or the environmental impact of products once they do become obsolete. Thus, in traditional production, products go from the "cradle"

4 Paul Hawken, Amory Lovings and L. Hunter Lovins, *Natural Capitalism* (New York: Little, Brown & Co., 1999).

(production in the factory) to the "grave" (disposal in a landfill or incineration).[5] An excellent example is the cell phone. Cell phones are manufactured to be disposed of every few years, as new ones are introduced into the market. Because they are so small, cell phones are often not recycled. Instead, they end up in enormous disposal yards where billions of cell phones, computers, and other old electrical technologies are buried, as if they will simply disappear. In fact, these "e-waste" sites are highly toxic in places like China; they are filled with lead and other poisonous metals, causing some of the highest cancer rates on the planet.[6]

Traditional manufacturing can be termed "monocultural"—one size fits all. We produce homogenous products that are the same everywhere despite differing requirements of diverse bio-regions. A good example of this kind of "monocultural product" is the house. Most houses are built in the same manner: a site is cleared, a house is constructed, and landscaping is added after the house is built. The problem is that different regions require ecologically distinct residential designs due to their unique climates, geographies, and resources available for building. Yet most developers in the United States use "one size fits all" housing prototypes for suburban housing construction.

A "cradle to cradle" approach to manufacturing housing would emphasize use of local materials, sensitivity to local geography, and "eco-efficiency," thus creating a range of housing prototypes for different eco-regions. This approach would also employ the use of local flora and fauna to ensure that landscaping is appropriate to its bio-region. Finally, it would take into account natural energy flows, so that designs in certain regions might feature wind towers or solar energy panels.

---

5 This section is derived from the excellent book, William McDonough and Michael Braungart, *Cradle to Cradle: Remaking the Way We Make Things*. New York: North Point Press, 2002.

6 In November 2008, CBS's "Sixty Minutes" program shocked the world with its video essay on this phenomenon. See CBS News, "The Electronic Wasteland" November 9, 2008. http://www.cbsnews.com/video/watch/?id=4586903n. See also Elizabeth Grossman, *High Tech Trash*. London: Island Press/Shearwater, 2007.

Fig. 8.4 House with photovoltaic and thermodynamic solar energy roof panels

Ecological Accounting

Ecological Design, published in 1996, argues that the environmental crisis is a design crisis.[7] In effect, the challenge for improving environmental sustainability is to better account for how things are made, how buildings are constructed, and how landscapes are used. We need to pay greater attention to place, and to the connections between nature and geometry. Human flows must be matched to nature's geometries across different scales. We need to emphasize the connection between organism and habitat (nature).

The accompanying chart offers an example of ecological accounting. It tracks the different energy flows created by a new building in San Francisco, showing the sources of electricity, garbage collection, natural gas, recycling, sewage, and water. It shifts our focus to the ecological footprint of buildings and their relationship to the surrounding region.

7 Sim Van Der Ryn and Stuart Cowan, *Ecological Design* (Washington, D.C.: Island Press, 1996).

FLOW	ACCOUNTING
ELECTRICITY	Canada, 25% Fuel plants, SF, 25% Nuclear (S.L. Obispo), 17% geothermal, 8% wind, solar power, Death Valley, 17% hydroelectric, Sierras, 8%
GARBAGE	1200 tons of trash" --transfer station (south of SF) --100 large trucks carry --to Altamont Landfill (70 mi. east), 1,528 acres --biogenic recovery facility, produce gas for 6000 homes
NATURAL GAS	California gas fields 11% Alberta gas fields 51% southwest gas fields 38%
RECYCLING	--used glass, melted, to Oakland --plastic bottles, tucked to Chino --aluminum, shipped to Texas --paper/cardboard, to Pacific Rim for recycling
SEWAGE	Oceanside Treatment Plant = 22 million gallons effluent, dumped 4 ½ mi. into Pacific Ocean North Point = 60 million gallons into SF bay
WATER	snowmelt in High Sierra, to regional reservoirs, pipelines, to municipal reservoirs

Source: Sim Van Der Ryn and Stuart Cowan, *Ecological Design* (Washington, D.C.: Island Press, 1996).

Future Strategies

This chapter raised a set of concerns for less developed nations and highlighted the challenges they face for economic development amidst burgeoning urban growth. These points are summarized here as an indication of future strategies for sustainable growth and the effective management of natural and cultural resources:

- *Restore degraded land.* Land should be catalogued and where compromised (with toxic waste, erosion, siltation, flooding, and so forth.), steps should be taken to create restoration policies.
- *Make cities fit their bio-regions.* Cities like Rio de Janeiro grew because of their natural surroundings, but then began to destroy those same natural features. Cities have to re-take control of the environment by carefully analyzing land, water, air, and vegetation and making sure these elements are protected from the pressures of population growth, industrialization, and expansion.
- *Balance development with nature.* Governments should abandon "one size fits all" development and begin to think in terms of development (buildings, infrastructure) tied to natural ecology.
- *Contain urban sprawl.* Urban development that spreads out makes it harder to move people efficiently. Governments need to rethink density of development, as well as the ecological footprint of the form of urbanized regions.
- *Optimize energy performance.* Improve energy efficiency so that supplies can serve the growing world population. Shift toward a paradigm that links energy more closely with urban and building form.
- *Rethink production methods.* Natural capitalism argues that the economy can become more efficient and less ecologically harmful if steps are taken to radically rethink production, technology, and the relationship between producers and consumers.
- *Respect community and the cultural landscape.* Local ecology and culture should be primary and essential factors in development, rather than being perceived as impediments or constraints that need to be managed or eliminated.

- *Value social equity.* Less developed nations have millions of people living in poverty, with no access to safe, affordable housing, clean water, sanitation, or other basic services National city-building and planning policies must craft land management and affordable housing policies to meet the current and future needs of these marginalized populations.

Credits

Figure 1.1. "Percent Poverty World Map" © Dwrcan (CC BY 3.0) at http://commons.wikimedia.org/wiki/File:Percent_poverty_world_map.png. • **Figure 1.2.** "Vista de la plaza Interior del Museo" © Arkhael (CC BY-SA 3.0) at http://commons.wikimedia.org/wiki/File:Vista_de_la_plaza_Interior_del_Museo.jpg. • **Figure 1.3.** "Favela-Nova Friburgo" © Nate Cull (CC BY-SA 2.0) at http://www.flickr.com/photos/natecull/3647271/. • **Figure 1.4.** "BlankMap-World-v2 " © Criostóir (CC BY-SA 3.0) at http://commons.wikimedia.org/wiki/File:BlankMap-World-v2.png • **Figure 1.5.** "Fruits bogota01" © Louise Wolff (CC BY-SA 3.0) at http://commons.wikimedia.org/wiki/File:Fruits_bogota01.jpg • **Figure 1.6.** "Slums of Ecuador Guayaquil" © Michael Shick (CC BY 2.0) at http://www.flickr.com/photos/84387582@N00/4114523591 • **Figure 1.7.** "Buenos Aires - San Nicolás - Avenida Corrientes - 200807a" © Mitchell Zappa (CC BY-SA 2.0) at http://www.flickr.com/photos/41894166489@N01/2730617928/ • **Planet Green logo.** Copyright in the Public Domain. • **Figure 2.1.** "Itahuania - Août 2008" © Martin St-Amant (CC BY-SA 3.0) at http://commons.wikimedia.org/wiki/File:7_-_Itahuania_-_Ao%C3%BBt_2008.JPG • **Figure 2.2.** "Cerro laguna verde atacama chile" © Gerard Prins (CC BY-SA 3.0) at http://commons.wikimedia.org/wiki/File:Cerro_laguna_verde_atacama_chile.jpg • **Figure 2.3.** "PlantacaodeSoja" © Tiago Fioreze (CC BY-SA 3.0) at http://commons.wikimedia.org/wiki/File:PlantacaodeSoja.JPG • **Figure 2.4.** Copyright in the Public Domain. • **Figure 2.5.** Copyright in the Public Domain. • **Figure 2.6.** Copyright in the Public Domain. • **Figure 2.7.** "COLLECTIE TROPENMUSEUM Het aftappen van een rubberboom in de omgeving van Banyuwangi TMnr 60052158" © Tropenmuseum of the Royal Tropical Institute (KIT) (CC BY-SA 3.0) at http://commons.wikimedia.org/wiki/File:COLLECTIE_TROPENMUSEUM_Het_aftappen_van_een_rubberboom_in_de_omgeving_van_Banyuwangi_TMnr_60052158.jpg • **Figure 2.8.** "Chico Mendes" © Yone Fernandes (CC BY 3.0 BR) at http://commons.wikimedia.org/wiki/File:Chico_Mendes.jpg • **Figure 2.9.** "Zócalo a Oaxaca" © Nsaum75 (CC BY-SA 3.0) at http://en.wikipedia.org/wiki/File:Z%C3%B3calo_a_Oaxaca.jpg • **Figure 2.10.** "DirkvdM canopy walk" © Dirk van der Made (CC BY-SA 3.0) at http://commons.wikimedia.org/wiki/File:DirkvdM_canopy_walk.jpg • "Iphone 2" © Daniel Zanetti (CC BY-SA 3.0) at http://commons.wikimedia.org/wiki/File:Iphone_2.jpg • **Figure 3.1.** "USA Texas Big Bend Santa Elena Canyon Bank" © Ralf Kiepert (CC BY-SA 3.0) at http://commons.wikimedia.org/wiki/File:USA_Texas_Big_Bend_Santa_Elena_Canyon_Bank_09-10-18.jpg • **Figure 3.2.** Copyright © 2011 Depositphotos/Stock Vector • **Figure 3.4.** Copyright in the Public Domain. • **Figure 3.6.** "Clothes hanging to dry" © Christian Haugen (CC BY 2.0) at http://www.flickr.com/photos/34073237@N04/3274224306 • **Figure 3.7.** Copyright in the Public Domain. • **Figure 3.8.** "Berazategui-Caacupe-cocina" © Cargale (CC BY 3.0) at http://commons.wikimedia.org/wiki/File:Berazategui-Caacupe-cocina.jpg • **Figure 3.9.** Copyright in the Public Domain. • **Figure 3.11.** "Catedral de Huajuapan de León" © Gunnar Wolf (CC BY-SA 3.0) at http://commons.wikimedia.org/wiki/File:Catedral_de_Huajuapan_de_Le%C3%B3n.jpg • **Figure 3.12.** Copyright in the Public Domain. • **Figure 4.2.** Copyright in the Public Domain. • **Figure 4.3.** "Central Park during Autumn, NYC" © Ekabhishek (CC BY-SA 2.0) at http://www.flickr.com/photos/72098626@N00/4085346513/ • **Figure 4.4.** Copyright in the Public Domain. • **Figure 4.5.** "747Af747" © Philippe Noret (CC BY-SA 2.5) at http://commons.wikimedia.org/wiki/File:747Af747.jpg • **Figure 4.6.** "Brasilia - Plan" ©

Uri Rosenheck (CC BY-SA 3.0) at http://commons.wikimedia.org/wiki/File:Brasilia_-_Plan.JPG • **Figure 4.7.** Copyright in the Public Domain. • **Figure 4.9.** "Suburbia by David Shankbone" © David Shankbone (CC BY-SA 3.0) at http://commons.wikimedia.org/wiki/File:Suburbia_by_David_Shankbone.jpg • **Figure 4.11.** "Waterfront Park from Hawthorne Bridge" © Steve Morgan (CC BY-SA 3.0) at http://commons.wikimedia.org/wiki/File:Waterfront_Park_from_Hawthorne_Bridge.jpg • **Figure 4.12.** "Public Transit in Curitiba - Brazil" © Adam Jones (CC BY-SA 3.0) http://commons.wikimedia.org/wiki/File:Public_Transit_in_Curitiba_-_Brazil.jpg • **Figure 4.13.** "Curitiba skyline" © Henri Bergius (CC BY 2.0) at http://www.flickr.com/photos/bergie/315936124/ • **Figure 5.1.** "Plaza Cuatro Caminos" © Alx 91 (CC BY-SA 2.5) at http://commons.wikimedia.org/wiki/File%3APlaza_Cuatro_Caminos.jpg • **Figure 5.3.** Copyright in the Public Domain. • **Figure 5.4–5.** "Imagebysafa2" © Safainla (CC BY-SA 3.0) at https://commons.wikimedia.org/wiki/File:Imagebysafa2.jpg • **Figure 5.7.** "Acapulco nammer" © Nammer (CC BY 2.0) at http://commons.wikimedia.org/wiki/File:Acapulco_nammer.jpg • **Figure 5.9.** Copyright in the Public Domain. • **Figure 5.10.** "Rosarito Beach Hotel" © Cesar Bojorquez (CC BY 2.0) at http://www.flickr.com/photos/uncut/181708602/sizes/l/ • **Figure 6.1.** Copyright in the Public Domain. • **Figure 6.1a.** "Santa Maria Beach" © StarbucksFreak (CC BY-SA 3.0) at http://commons.wikimedia.org/wiki/File:Santa_Maria_Beach.jpg • **Figure 6.1b.** "South American Jungle SG" © Sascha Grabow (CC BY-SA 3.0) at http://commons.wikimedia.org/wiki/File:South_American_Jungle_SG.jpg • **Figure 6.3.** "Sra. Charo and the Qewar dollmakers" © Bebop1065 (CC BY-SA 3.0) at http://commons.wikimedia.org/wiki/File:Sra._Charo_and_the_Qewar_dollmakers.jpg • **Figure 6.4.** "6828 Puerto Ordaz" © Berrucomons (CC BY-SA 3.0) at http://commons.wikimedia.org/wiki/File:6828_Puerto_Ordaz.jpg • **Figure 6.5.** "Poblado de Borja" © Carlos Correa Loyola (CC BY 3.0 EC) at http://commons.wikimedia.org/wiki/File:Poblado_de_Borja.jpg • **Figure 6.6.** "Market in Pisac Peru" © AFiebrandt (CC BY-SA 3.0) at http://commons.wikimedia.org/wiki/File:Market_in_Pisac_Peru.JPG • **Figure 6.8.** "Chinchero market" © Bernard Gagnon (CC BY-SA 3.0) at http://commons.wikimedia.org/wiki/File:Chinchero_market.jpg • **Figure 7.2.** "Rocinha Favela Brazil Slums" © Alicia Nijdam (CC BY 2.0) at http://www.flickr.com/photos/24265317@N00/2361799355 • **Figure 7.3.** Copyright in the Public Domain. • **Figure 7.4.** "AerialViewMexicoCity" © Fidel Gonzalez (CC BY-SA 3.0) at http://commons.wikimedia.org/wiki/File:AerialViewMexicoCity.jpg • **Figure 7.5.** "SmogOverMexicoCity" © Fidel Gonzalez (CC BY-SA 3.0) at http://commons.wikimedia.org/wiki/File:SmogOverMexicoCity.jpg • **Figure 7.7.** "Colonia Atlampa Mexico DF" © Iván Martinez (CC BY-SA 3.0) at http://commons.wikimedia.org/wiki/File:Colonia_Atlampa_Mexico_DF_-_1.JPG • **Figure 7.8.** Copyright in the Public Domain. • **Figure 7.9.** "Caracas - Parque del Este" © Márcio Cabral de Moura (CC BY 2.0) at http://www.flickr.com/photos/mcdemoura/2318001539/ • **Figure 7.14.** "Santa Cruz fort and the opening of the Guanabara bay" © Eurico Zimbres (CC BY-SA 3.0) at http://commons.wikimedia.org/wiki/File:Santa_Cruz_fort_and_the_opening_of_the_Guanabara_bay_.JPG • **Figure 7.15.** "Favelas-Caxias" © Ricardo Frantz (CC BY 3.0) at http://commons.wikimedia.org/wiki/File:Favelas-Caxias.jpg • **Figure 8.1.** Copyright in the Public Domain. • **Figure 8.2.** Copyright in the Public Domain. • **Figure 8.3b.** "Velro Hooks" © Alexander Klink (CC BY 3.0) at http://commons.wikimedia.org/wiki/File:Velcro_Hooks.jpg • **Figure 8.4.** "Bur Macro BlackBg" © Zephyris (CC BY-SA 3.0) at http://en.wikipedia.org/wiki/File:Bur_Macro_BlackBg.jpg • **Figure 8.9.** Copyright in the Public Domain. • **Table, p. 104.** Sim Van Der Ryn and Stuart Cowan, Ecological Design. Copyright © 1996 by Island Press. Reprinted with permission. •

www.ingramcontent.com/pod-product-compliance
Lightning Source LLC
Chambersburg PA
CBHW071251180926
51613CB00001B/8
9781516551682